What We Are: The Evolutionary Roots
of Our Future

Lonnie Aarssen

What We Are:
The Evolutionary Roots
of Our Future

 Springer

Lonnie Aarssen
Department of Biology
Queen's University
Kingston, ON, Canada

1st edition: © Author 2015

Cover art by Christiane Beauregard—http://www.christianebeauregard.com/

ISBN 978-3-031-05881-3   ISBN 978-3-031-05879-0   (eBook)
https://doi.org/10.1007/978-3-031-05879-0

This Springer imprint is published by the registered company Springer Nature Switzerland AG
The registered company address is: Gewerbestrasse 11, 6330 Cham, Switzerland

TO

WILLIAM CHARLES LAWRENCE

Prelude

This book is an update of my earlier book with a similar title: Aarssen (2015) *What Are We? Exploring the Evolutionary Roots of Our Future*. Chapter titles are the same here but with a new chapter added at the end. Many additional—including of course more recent—references have been added with associated expansion and further development of key concepts, ideas, and hypotheses supported by these references.

For many insightful discussions of topics in this book, I am grateful to my best friend and beloved partner in life, Janice, and to my many colleagues and students, too numerous to name. For their endless patience and resourcefulness, I also thank Catherine DeNoble for helping me to secure image copyright credits and attributions, and Shina Harshavardhan, Kenneth Teng, and the rest of the Springer Nature editing and production team for guiding this project to completion.

Preface

[From Aarssen L (2015) What are we? Exploring the Evolutionary Roots of Our Future. Queen's University, Kingston]

Paul Gauguin (1897) *D'où venons-nous? Que sommes-nous? Où allons-nous? (Where do we come from? Who are we? Where are we going?)*/Wikimedia Commons/Public Domain

Humans are fascinated with themselves. What are we? Do our lives mean something? Our obsession with these questions is why the arts and humanities exist. They have given us a rich and bewildering variety of aesthetics, symbolisms, ideologies, narratives, melodies, myths, meditations, beliefs, philosophies, customs, gods, rituals, entertainments, and institutions that have come and gone across recorded history. As acclaimed writer John Updike put it:

> To be human is to be in the tense condition of a death-foreseeing consciously libidinous animal. No other earthly creature suffers such a capacity for thought, such a complexity of envisioned but frustrated possibilities, such a troubling ability to question the tribal and biological imperatives. So conflicted and ingenious a creature makes an endlessly interesting focus for the meditations of fiction. (Updike 2000, *The Tried and the Treowe*)

The arts and humanities, accordingly, have always thrived from pluralism of interpretation for the human experience, particularly one's mental life—the "inner self." In other words, the "What are we?" question must remain unanswered—celebrated as an enduring and revered mystery. After all, humans are fascinated with not just themselves but also with mystery and novelty, and with stories and surprises—especially about themselves. These are enjoyed daily by billions of people watching films and theater, reading novels and other literature, attending concerts and carnivals, and studying, worshiping, and otherwise indulging in countless other products of culture. We humans are creatures of emotion, impulsively drawn to come back again and again to indulge in more. Perhaps it reassures us that we are alive.

Nevertheless, answers—or where to discover them—have indeed been found. Science has, with wide consensus in recent decades, given us a very clear and certain perspective of what we are: We are an animal among many millions of others, the vast majority of which have long been extinct—a species that is only about 300,000 years old, but descended from a long lineage, most of which was not human. This deep history of our origin and our placement and function in the biosphere through time—as for all other species—unfolded because of Darwinian evolution by natural selection. And this discovery has given us what the arts and humanities never could, and never aspired to find: vital insight into how and why human nature, social life, and culture have come to be what they are, and so profoundly different from other species.

Insights from evolution, however, particularly about what humans are, have never been met with enthusiasm from the general public, nor from many professionals. Unfortunately, they are commonly misinterpreted as deliberately threatening, insulting, or even sinister. Darwinism has always been an uncomfortable truth—unintentionally but bluntly challenging the heartfelt beliefs and sensibilities of many good and well-intentioned people. And Darwinism also inadvertently calls upon the arts and humanities—and much of the social sciences—to re-examine what *they* are, and to imagine anew what they have potential to be, now that many of the ageless curiosities and perplexities of the human condition are being answered, with growing precision, by evolutionary biology.

This book is a brief survey of this evolutionary interpretation of what we are, and where we came from—drawing from, and integrating across, several fields of study. It includes emphasis on some recent discoveries, and also explores some new ideas and hypotheses, pointing to inspiration for future research, including with potential to develop more common ground for the life sciences to share together with the arts, humanities, and social sciences.

But the primary goal of this book is much more profound and far-reaching. Arriving at a concise and broadly public understanding of what we are has never been more urgent—because of *what we have done*. Over the short time span of human evolution, *Homo sapiens* have become the hyper-dominant animal on the planet, recklessly overharvesting Earth's resources, obliterating other species, and degrading or destroying the ecosystem services upon which human wellness depends. The scale and impact of these effects have multiplied severalfold over just the few decades of my lifetime, and we are now faced with some very alarming questions—and a growing number of frightening certainties—about where our species is headed. I am not normally inclined to be a doomsayer, and I am as weary as any from all of the bad news. Nevertheless, indications from several reliable sources all point to an impending collapse of "business as usual" for human civilization. And very few people are paying any attention.

In response to this crisis, I developed a course at Queen's University in 2008, called *Evolution and Human Affairs*. The goal was, and remains, not just to help students find greater awareness of the converging catastrophes of modern civilization but also to find a deeper understanding of how our evolutionary roots—by shaping what we are—have brought us to this critical point in the history of humanity. This book is an account of what my students and I are learning together about our human journey. Some of it points to hope for the future—but some of it, not so much. Our greatest limitation may be that we don't really know ourselves very well at all.

Even more concerning is that we may actually prefer not to know ourselves too well. As poet T.S. Eliot mused, "… humankind cannot bear very much reality" (Eliot 1943, No. 1 of *Four Quartets*). In this book, we explore how it is easier to confront what we are by discovering how we got that way, and that this in turn prepares us for deeper insight into where we and our planet are likely to be headed. Only recently have we acquired the tools of science needed to study and discover these things. From the arts and humanities, we have enjoyed a long and enchanting history of wonder, introspection, and imagination about the human condition. And for the future, we can expect continuing enrichment from these pursuits. But as global citizens in the twenty-first century, we can no longer afford to remain content with just the allure of intrigue, the excitement of serendipity, the charm of stories, the visions of mystics, the superstitions of theology, and the bliss of ignorance.

The critical question then is this: *Has our evolution, as a species, equipped us to respond effectively to the converging catastrophes of the twenty-first century?* It is my belief that students, and others who read this book, will be better equipped to answer this question, and thus to capture a glimpse of the evolutionary roots of our future, and so to participate in prescribing a way forward for the design of a new, more sustainable, and more humanistic model of civilization. Whether there is enough time left to do so, I am less certain.

Kingston, ON, Canada   Lonnie Aarssen
August 2015

Man has not only evolved; for better or worse, he is evolving. Our not very remote ancestors were animals, not men; the transition from animal to man is, on the evolutionary time scale, rather recent. But the newcomer, the human species proved fit when tested in the crucible of natural selection; this high fitness is a product of the genetic equipment which made culture possible. Has the development of culture nullified the genes? Nothing could be more false. Culture is built on a shifting genetic foundation. It is fairly generally admitted that genetic changes in the human species are influenced by culture. But many people are reluctant to credit that genetic changes may influence culture. The reluctance comes from an almost obsessive fear that biological influences on culture are somehow incompatible with democratic ideals; social sciences must be guarded against the encroachment of biology. ... But the estrangement must be overcome. Man's future inexorably depends on the interactions of biological and social forces. Understanding these forces and their interactions may, in the fullness of time, prove to be the main achievement of science.
—Theodosius Dobzhansky (1962) *Mankind Evolving*

Contents

About the Author

Lonnie Aarssen is Professor of Biology at Queen's University, Kingston, Canada. He has served on the editorial board of several academic journals and is the founding editor of the open-access journal *Ideas in Ecology and Evolution* published at Queen's University. He has authored or co-authored more than 170 peer-reviewed publications and has disseminated his work in Tedx talks and Science Animated videos. In his recent teaching and writing, he explores how evolutionary thinking can affect our understanding of our lives, our species, and our ability to share the planet with other species.

Chapter 1
What Have We Done?

… the impact of our race upon the environment has so increased in force that it is has changed in essence. … Our present combustion of fossil fuels threatens to change the chemistry of the globe's atmosphere as a whole, with consequences which we are only beginning to guess. With the population explosion, the carcinoma of planless urbanism, the now geological deposits of sewage and garbage, surely no creature other than man has ever managed to foul its nest in such short order. (White 1967)

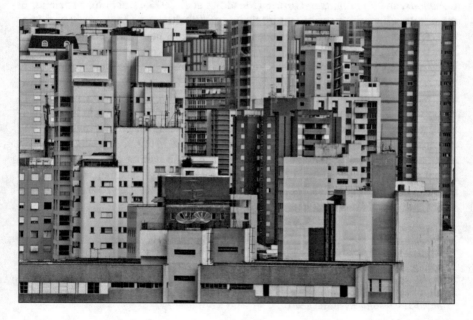

Paulisson Miura (2012) *Paisagem caótica (Chaotic Cityscape)* (https://www.flickr.com/photos/paulisson_miura/). (Used with permission)

Princeton professor Lynn White Jr. delivered the above diagnosis in a now famous lecture—'The Historical Roots of Our Ecological Crisis'—at the Washington

L. Aarssen, *What We Are: The Evolutionary Roots of Our Future*,
https://doi.org/10.1007/978-3-031-05879-0_1

meeting of the American Association for the Advancement of Science in 1966. More than half a century later, the nest of humanity is fouler still. Climate change is now the poster child of the environmental movement. Since 1988, its pace, causes, and consequences have been the subjects of intense study and lucid forecasting by the Intergovernmental Panel on Climate Change. Its mission has become one of the largest and most in-depth analyses of human impact on the planet ever organized, involving contributions from dozens of countries, hundreds of researchers and authors, and many thousands of peer reviewers. Each of its reports is more alarming than the previous one, and the most recent synthesis report (at the time of writing) (https://www.ipcc.ch/report/ar5/wg2/) is especially daunting, with increasingly certain predictions for dramatic and large-scale deterioration and loss of ecosystem services and human well-being over the course of this century (IPCC 2021).

Climate change of course is only one of several environmental disasters that are threatening human civilization as we know it. The alarm began sounding conspicuously in the 1960s, famously with Rachel Carson's (1962) *Silent Spring*, Paul Ehrlich's (1968) *The Population Bomb*, Garrett Hardin's (1968) *Tragedy of the Commons*, and *The Limits to Growth* (Meadows et al. 1972). But human societies in the developed world were caught up then in the excitement of the 'great acceleration' (Steffen et al. 2015) and its false promise of a prosperous future for humanity. New technologies and opportunities for economic growth began to emerge rapidly in the middle of the last century, bringing huge increases in agricultural productivity and extraction/harvesting rates of natural resources (oil, water, timber, fisheries) and, with this, a growing addiction to consumerism.

By the 1990s, many groups of scientists were trying hard to reign in this exuberance, with public calls for restraint and cautionary appeals for greater understanding of impact in the longer view, involving sharp increases in human population growth rate, carbon emissions, pollution, and loss of wildlife habitat and biodiversity on global scales. In 1992, about 1700 of the world's leading scientists, including the majority of Nobel laureates, issued the *World Scientists' Warning to Humanity*. The Introduction pulled no punches:

> *Human beings and the natural world are on a collision course. Human activities inflict harsh and often irreversible damage on the environment and on critical resources. If not checked, many of our current practices put at serious risk the future that we wish for human society and the plant and animal kingdoms, and may so alter the living world that it will be unable to sustain life in the manner that we know. Fundamental changes are urgent if we are to avoid the collision our present course will bring about (http://www.ucsusa.org/about/1992-world-scientists.html).*

In the same year, a similar appeal was issued jointly by the Royal Society and the National Academy of Sciences (reproduced in the journal *Population and Development Review* 18: 375–378). This was followed in 1994 by a Science Summit on World Population: A Joint Statement by 58 of the World's Scientific Academies (*Population and Development Review* 20: 233–238). From the latter: 'In our judgement, humanity's ability to deal successfully with its social, economic, and environmental problems will require the achievement of zero population growth within the lifetime of our children'.

Deaf Ears

While these warnings continued to fall mostly on deaf ears, many groups of scientists and other experts ramped up their efforts vigorously in the new millennium, issuing several detailed reports documenting the imperilled trajectory of human civilization:

The Living Planet Reports (beginning in 2000):

(http://www.footprintnetwork.org/en/index.php/GFN/page/living_planet_report2/).

Global Environment Outlook Reports (beginning in 2000):

(http://www.unep.org/geo/).

Millennium Ecosystem Assessment (2005):

(http://www.millenniumassessment.org/en/index.html).

Earth System Science for Global Sustainability – Grand Challenges (Reid et al. 2010):

We know enough to state with a high degree of scientific confidence that without action to mitigate drivers of dangerous global change and enhance societal resilience, humanity has reached a point in history at which changes in climate, hydrological cycles, food systems, sea level, biodiversity, ecosystem services and other factors will undermine development prospects and cause significant human suffering associated with hunger, disease, migration and poverty. If unchecked or unmitigated, these changes will retard or reverse progress towards broadly shared economic, social, environmental and developmental goals.

State of the Planet Declaration (Brito and Stafford-Smith 2012):

Research now demonstrates that the continued functioning of the Earth system as it has supported the well-being human civilization in recent centuries is at risk. Without urgent action, we could face threats to water, food, biodiversity and other critical resources: these threats risk intensifying economic, ecological and social crises, creating the potential for a humanitarian emergency on a global scale.

Most of humanity has continued not to listen or not to care (Tollefson and Gilbert 2012). Remaining undeterred, a 2013 'Consensus Statement from Global Scientists', on 'Maintaining Humanity's Life Support Systems in the 21st Century', sounds eerily like the World Scientists' Warning to Humanity from two decades earlier:

Earth is rapidly approaching a tipping point. Human impacts are causing alarming levels of harm to our planet. As scientists who study the interactions of people with the rest of the biosphere using a wide range of approaches, we agree that the evidence that humans are damaging their ecological life-support systems is overwhelming. We further agree that, based on the best scientific information available, human quality of life will suffer substantial degradation by the year 2050 if we continue on our current path (https://consensus-foraction.stanford.edu/see-scientific-consensus/consensus_english.pdf).

And 25 years after the first World Scientists' Warning to Humanity, the Alliance of World Scientists tried again with 'A Second Notice' (Ripple et al. 2017), this time with 15,364 signatories from 184 countries and a presentation of time series data with predictably staggering implications:

By failing to adequately limit population growth, reassess the role of an economy rooted in growth, reduce greenhouse gases, incentivize renewable energy, protect habitat, restore ecosystems, curb pollution, halt defaunation, and constrain invasive alien species, humanity is not taking the urgent steps needed to safeguard our imperilled biosphere.

Similarly, more than 40 years after the iconic report, *The Limits to Growth* (Meadows et al. 1972), an update of that analysis gives sharp warning of a looming catastrophe: 'There is high risk for pushing the Earth's life supporting systems beyond irreversible trigger-points by 2050' (von Weizsacker and Wijkman 2018). Yet still, as I prepare this opening chapter, another plea from leading scholars in environmental and conservation science has landed on my desk, predicting a 'ghastly future' for humanity (Bradshaw et al. 2021).

Sources and Sinks, Footprints and Capacities

Much has been written about the details of our environmental crisis, including in the reports cited above. Only a brief summary is needed here. Modern civilization depends on the use of materials and energy. It thus requires resource inputs ('sources') from nature and the physical environment, and it necessarily produces waste outputs ('sinks') back into the environment. As with any source/sink relationship, it can be sustained only if sources are not overly depleted and if sinks are not overly filled, and this depends critically on the rate of resource use. In other words, extracted resources must be regenerated, and accumulated wastes must be recycled (thus replenishing the 'sources') or they must be broken down into harmless components. If this fails perpetually, then vital ecosystems services are lost and civilization collapses.

The degradation of ecosystem services is now critical and is in large part a consequence of the rapid growth in human population size, particularly since the beginning of the last century. The earth is now more than full of *Homo sapiens*. But it is fuller in some places than in others. In poor, overcrowded countries, the Human Development Index (quality of life measured by life expectancy, literacy and education, and per capita GDP) is unacceptably low, but the demands on ecosystem services–the 'ecological footprint'–are relatively small (Living Planet Report 2020). In contrast, in more developed countries, Human Development Index is higher of course, not just because population density is lower but also because of greater affluence supported by higher rates of consumption per capita and availability of the modern technologies and economies required to access the energy production and resource extraction rates (with their attendant emissions/wastes) that enable this affluence. The latter generally imposes a disproportionately large negative impact on ecosystem services (a large ecological footprint), even though population size is relatively small. A summary of the main variables is given by what is sometimes referred to as the 'IPAT' equation:

$$\text{IMPACT of humans on the degradation of ecosystem services}$$
$$= \text{POPULATION size} \left(\text{affecting total consumption and waste production}\right)$$
$$\times \text{AFFLUENCE} \left(\text{consumption per capita}\right)$$
$$\times \text{TECHNOLOGY} \left(\text{allowing affluence; but with negative effects, e.g. pollution}\right)$$

The continuing destruction of natural habitat for wildlife has now set in motion the earth's sixth mass extinction event (Dirzo et al. 2014, Gilbert 2018), affecting everything from the smallest invertebrates to the largest mammals. The overall impact of our species on the planet is now more than 50% greater than what nature can renew. In other words, it would take 1.54 earths to sustainably meet the demands that humanity currently makes on nature (Living Planet Report 2020). With a global population size now over 7.5 billion and rising, it is clearly not possible to achieve and maintain an acceptable Human Development Index without imposing an ecological footprint that exceeds the global capacity of ecosystem services to regenerate. And most of the less developed world is of course aiming to enjoy – as soon as possible – the consumerism and affluence long enjoyed by more developed nations, and many are managing to do so. Civilization cannot continue with 'business as usual' for much longer.

Looking Ahead by Looking Back

Our species is faced with a profound reckoning for what it has done. The urgency and far-reaching implications of the above warnings and predictions have made major news headlines, periodically, over the past several decades. Yet most in the general public (and hence most governments) have so far reacted with little, short-lived, or no significant alarm. Humans seem generally ill-equipped to respond effectively to the converging catastrophes of the twenty-first century. Clearly, the problem is *us*. The great challenge for humanity then is to transform 'us' from being the problem, into being the solution. But if we are to become the 'solution', we need to first understand how/why we became the 'problem'. In this book we will examine how our collapsing civilization as well as our dysfunctional reaction to it are products of our own biocultural evolutionary history. In other words, they are products of human motivations that generously rewarded the reproductive success of our ancestors – motivations therefore that define what we are today: our behaviours, social lives, and cultures.

In the last two chapters, we explore how a deep understanding of these evolutionary roots of what we are is critical for the next 'episode' of the human journey: designing a new and improved project of civilization for our descendants. Their future lies in what we are. And what we are is from the past.

If uncomfortable truths are out there, we should seek them and face them like intellectual adults, rather than eschew open-minded inquiry or fabricating philosophical theories whose only virtue is the promise of providing the soothing news that all our heartfelt beliefs are true. Joyce (2006).

References

Bradshaw CJA, Ehrlich PR, Beattie A, Ceballos G, Crist E, Diamond J, Dirzo R, Ehrlich AH, Harte J, Harte ME, Pyke G, Raven PH, Ripple WJ, Saltré F, Turnbull C, Wackernagel M, Blumstein DT (2021) Underestimating the challenges of avoiding a ghastly future. Front Conserv Sci 1:615419. https://doi.org/10.3389/fcosc.2020.615419

Brito L, Stafford-Smith M (2012) State of the planet declaration. Planet under pressure: new knowledge towards solutions conference, London, 26–29 Mar 2012. Available at: http://www.igbp.net/download/18.6b007aff13cb59eff6411bbc/1376383161076/SotP_declaration-A5-for_web.pdf

Carson RL (1962) Silent spring. Houghtoon Mifflin, New York

Dirzo R, Young HS, Galetti M, Ceballos G, Isaac NJB, Collen B (2014) Defaunation in the anthropocene. Science 345:401. https://doi.org/10.1126/science.1251817

Ehrlich P (1968) The population bomb. Ballantine Books, New York

Gilbert N (2018) Top UN panel paints bleak picture of world's ecosystems. Nature, News 27 Mar 2018. https://www.nature.com/articles/d41586-018-03891-1

Hardin G (1968) The tragedy of the commons. Science 162:1243–1248. http://www.sciencemag.org/content/162/3859/1243.full.pdf

IPCC (2021) Climate change 2021: the physical science basis. Contribution of working group I to the sixth assessment report of the intergovernmental panel on climate change. In: Masson-Delmotte V, Zhai P, Pirani A, Connors SL, Péan C, Berger S, Caud N, Chen Y, Goldfarb L, Gomis MI, Huang M, Leitzell K, Lonnoy E, Matthews JBR, Maycock TK, Waterfield T, Yelekçi O, Yu R, Zhou B (eds). Cambridge University Press, In Press. https://www.ipcc.ch/report/ar6/wg1/#FullReport

Joyce R (2006) The evolution of morality. MIT Press, Cambridge

Living Planet Report. (2020) https://www.footprintnetwork.org/content/uploads/2020/09/LPR2020-Full-report-lo-res.pdf

Meadows DH, Meadows DL, Randers J, Behrens WW III (1972) The limits to growth: a report for the club of Rome's project on the predicament of mankind. Universe Books, New York

Reid WV, Chen D, Goldfarb L, Hackmann H, Lee YT, Mokhele K, Ostrom E, Raivio K, Rockström J, Schellnhuber HJ, Whyte A (2010) Earth system science for global sustainability: grand challenges. Science 330:916–917

Ripple WJ, Wolf C, Newsome TM, Galetti M, Alamgir M, Crist E, Mahmoud MI, Laurance WF (2017) World scientists' warning to humanity: a second notice. Bio Science 67:1026–1028. https://doi.org/10.1093/biosci/bix125

Steffen W, Broadgate W, Deutsch L, Gaffney O, Ludwig C (2015) The trajectory of the anthropocene: the great acceleration. Anthropocene Rev 2:81–98

Tollefson J, Gilbert N (2012) Rio Report Card. Nature 486:20–23

von Weizsacker E, Wijkman A (2018) Come on! capitalism, short-termism, population and the destruction of the planet: a report to the club of Rome. Springer, New York

White L (1967) The historical roots of our ecologic crisis. Science 155:1203–1207

Chapter 2
A Primer on Evolutionary Roots

To arrive at where we want to be, we need to take full account of where we came from and of what we are. (Stamos 2008)

Glenn Jones (2010) *Evolution* (http://glennz.tumblr.com/post/457492074/evolution). (Used with permission)

Darwinian evolution is the underlying theme of this book. Its main principles are not complicated, and only a summary is required here for the reader (Box 2.1). Two concepts, in particular, are of central importance for brief review: natural selection and evolutionary fitness.

L. Aarssen, *What We Are: The Evolutionary Roots of Our Future*,
https://doi.org/10.1007/978-3-031-05879-0_2

Natural Selection

Natural selection in action is represented by the failure of certain individuals to leave descendants through reproduction or to leave as many descendants as other individuals. Failure occurs because these certain individuals lack certain phenotypic traits or combinations of particular traits, e.g. affecting ability to avoid a predator or to attract a mate. Natural selection then 'acts' because of inferior traits, where the actions or 'forces of natural selection' are represented by things like the mortality imposed by a predator or the mate-choosing preferences/decisions imposed by a potential mate. Natural selection thus 'favours' those traits that promote descendant production in the individuals that bear them, by *disfavouring* those traits that do not or that promote fewer descendants. More specifically, if a trait is 'favoured' by natural selection, this means that its relative frequency is maintained or increased in subsequent generations *because* 'disfavoured' traits (affected by the same force of natural selection) decrease in frequency or disappear. This of course requires that the trait be heritable, i.e. informed by effects of genes transmitted by sex (Box 2.1). [This contrasts with cultural selection in cultural evolution (see Chap. 6), where some behavioural/ cultural phenotypes (especially in humans) are informed by effects of social learning (and need not also be informed by effects of genes) and are transmitted to subsequent generations by copying and communication (oral and written)].

Box 2.1: Key Principles of Darwinian Evolution by Natural Selection

Biological evolution refers to change in the genetic makeup of a population or species

It happens because:

(1) Individuals display *variation in certain traits*; the variants are called phenotypes.

(2) Some of this variation results from *differences in genes*; the variants are then called genotypes (A and B); and so the associated phenotypes are *heritable*, i.e. they can be passed on from one generation to the next.

(3) Some of these heritable traits ▮◖◗ affect *differential success in reproduction*, either directly (through effects on success with sex) or indirectly (through effects on growth and survival to reproductive maturity).

Traits that confer less success are said to be disfavoured by *natural selection*, while those conferring greater success are favoured.

Some genotypes therefore are more successful than others in leaving descendants, and thus transmitting copies of their genes to future generations; these genes therefore have higher *evolutionary fitness* under the prevailing environmental conditions.

The result is a change over time in the frequency of different genes/genotypes, and hence their associated phenotypes — i.e. *biological evolution*.

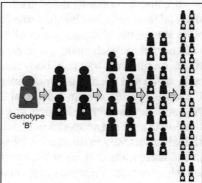

Genotype 'B'

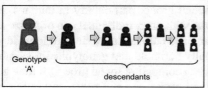

Genotype 'A'

descendants

Where does genetic variation come from, and how does it persist within a population or species?

(1) Genetic variation ultimately comes from *MUTATION* — changes in the structure of genes caused by environmental factors (e.g. chemicals, radiation) or by random mistakes in replication and re-assortment of genes during cell division and gamete (sperm and egg) formation. Many of these changes are detrimental, many are neutral, and some are beneficial in their effects on reproductive success; hence several different forms (alleles) of the same gene ◖◗ usually coexist within a population or species, and there is on-going formation of new forms.

(2) *SEXUAL REPRODUCTION* causes mixing of the different genes and their alleles (coming from a mother and a father) that affect different traits. Some of this mixing has secondary effects because the presence of one gene, or form of a gene, can sometimes affect the expression of another; hence several different genotypes, and hence phenotypes usually coexist within a population or species, and are generated anew within each generation.

(3) *GENOTYPE X ENVIRONMENT INTERACTION* — Some environments disfavour the reproductive success of certain genes/genotypes while favouring others, some genes can be turned 'on' or 'off' because of different environments, some genes affect more than one trait, and traits are commonly affected by more than one gene; Hence, since environmental conditions are usually heterogenous, in space and in time, several different genotypes, and different forms of the same gene can coexist within a population or species because each can be favoured under a different set of environmental conditions, or depending on which other genes/alleles reside in the same individual.

When a trait is favoured by natural selection, and its expression is influenced by particular genes, then their frequencies are also increased (or decreased) in subsequent generations. Hence, these particular genes are also said to be indirectly 'favoured' (or disfavoured) as a consequence of natural selection—and other resident genes in the same individual may 'go along for the ride', i.e. also with elevated copying and transmission success (even though the traits that they inform may not be particularly favoured or disfavoured by selection). Similarly, because the individuals bearing copies of the favoured genes leave more descendants, these individuals are also said to be indirectly 'favoured' as a consequence of natural selection. Importantly, this does not mean that natural selection *acts* on individuals or genes. Again, traits are the principal target of natural selection, and individuals and their resident genes can, only consequentially, be favoured (or disfavoured) *indirectly*. Thus, for example, offspring production for two individuals (and their resident genes) can be equally favoured indirectly by natural selection but for different reasons—because natural selection can act on (i.e. favour) different traits in each individual, and this could involve the same force or different forces of natural selection. In other words, two different forces of natural selection, e.g. mortality imposed by a predator and mate-choosing preferences/decisions imposed by a potential mate, could act on the same trait (e.g. body size), or they might act on different traits (e.g. running speed and courtship display, respectively).

For some species, certain individual traits involve social skills that can promote individual benefits from belonging to a social group, particularly in humans, e.g. inclination to seek out cooperation networks with other individuals, to organize and participate effectively in division of labour, or to cultivate allegiance to local ideology. And these traits may be at least partially heritable and genotypically variable. When this genotypic variation occurs *between* competing social groups, it may generate differential group 'success'—measured by its size and prosperity over time, i.e. number of resident individuals and their average quality of life. A social group, therefore, is sometimes also said to be 'favoured' indirectly (because other groups are disfavoured) as a consequence of natural selection and/or cultural selection acting on (i.e. disfavouring) traits expressed in relatively high frequency by resident individuals within the disfavoured group(s). This is sometimes referred to as 'group selection'—but to emphasize again, the direct targets of natural selection are always individual traits, not individuals or groups of individuals; groups can be favoured only indirectly. In other words, the propensities that define differences in group success are a function of between-group variation in the dispositions of resident individuals, i.e. the phenotypic expressions of heritable traits (subject to effects of natural selection) that characterize individual minds, e.g. involving variation in degrees of (and features that characterize) local knowledge, traditions, and prosocial motivations, including things like moral and empathic instincts and the 'need to belong' or to feel acceptance by others. We will return to these themes in later chapters.

In Chap. 6 we will explore how effects of natural selection and cultural selection are commonly blended, so that biological evolution influences cultural evolution (Ridley 2003) and vice versa. The result—*biocultural evolution*—accounts in large measure for how our species has become the most prolific animal on the planet and why we are now faced with the prospects of catastrophic collapse of civilization sometime in this century.

Evolutionary Fitness

Evolutionary fitness is commonly referred to as a property of individuals, i.e. those who leave more descendants are commonly said to have higher fitness. But this is understood only as an estimate of fitness. Biological evolution is defined in terms of change in gene frequencies, and hence change in the genetic constitution of a population, which results when different genes or forms of genes (alleles) are copied and transmitted (by reproduction) into future generations at different frequencies. Evolutionary fitness, therefore, really refers to genetic fitness. In other words, *it applies strictly to genes*—defined in terms of the number (or relative number) of gene copies residing in future generations (or in later generations compared with earlier generations).[1]

Genes with high fitness therefore are able to promote their own representation in future generations, and so—anthropomorphically—they can be thought of as 'selfish' in their own right (Dawkins 1976; Agren 2021). Some of the most prolific genes inform the expression of traits (including many human behaviours) in ways that are beneficial not just for their own transmission success but also for the transmission success of copies of those genes residing in other individuals. This transmission success is referred to as 'inclusive fitness' and explains, for example, why individuals (in every culture) are more inclined to be helpful to their close kin than to non-kin.

Importantly, however, neither genes nor the particular traits that they inform can propel themselves into future generations. Only whole individuals can do this through the production of offspring, usually involving sex (and always so in humans and most other animals). Because of this, the number of gene copies transmitted to future generations is generally correlated with the number of descendants that one leaves from direct lineage (which is why the latter is a convenient estimate of

---

[1] Gene frequencies may also change within a local population when individuals (and their resident genes) immigrate or emigrate (referred to as 'gene flow') or when a population becomes relatively small, and thus with relatively few individuals to carry gene/allele copies; by chance alone, therefore, most or all copies of certain genes/alleles may be lost from the population (referred to as 'genetic drift').

evolutionary fitness). But this is predictable only in the short term; it becomes less and less certain in more distant generations. All descendants carry gene copies from some ancestors. But from some (or even many) other distant ancestors, some (or even all) descendants may carry not a single gene copy. This is why evolutionary fitness is a property only of genes—not individuals or groups of individuals. An excerpt from Dawkins (2004) illustrates:

> *Every time an individual has a child exactly half of his genes go into that child. Every time he has a grandchild, a quarter of his genes on average go into that child. Unlike the first generation offspring where the percentage contribution is exact, the figure for each grandchild is statistical. It could be more than a quarter, it could be less. Half your genes come from your father, half from your mother. When you make a child, you put half of your genes into her. But which half of your genes do you give to the child? On average they will be drawn equally from the ones you originally got from the child's grandfather and the ones you originally got from the child's grandmother. But by chance you could happen to give all your mother's genes to your child, and none of your father's. In this case, your father would have given no genes to his grandchild. Of course, such a scenario is highly unlikely, but as we go down to more distant descendants, total non-contribution of genes becomes more possible. On average you can expect one-eighth of your genes to end up in each great-grandchild, one-sixteenth in each great-great-grandchild, but it could be more or it could be less. And so on until the likelihood of a literally zero contribution to a given descendant becomes significant.*

Depths of Darwinism

Dennett (1995) likened Darwinism to a 'universal acid'—so powerful that it has potential to transform virtually anything that it is applied to (e.g. see Brinkworth and Weinert 2012). Yet even a century and a half after the publication of the *Origin of Species*, Darwinism remains relatively unpopular as a cultural worldview, with many enduring misconceptions (Le Page 2008), and despite conspicuous evidence for recent effects of natural selection in everyday life. Obvious examples include the evolution of antibiotic resistance and pesticide resistance (Palumbi 2001; Antonovics et al. 2007) and the long history of animal domestication, where the effects of artificial selection by humans illustrate an exact parallel to the mechanism of natural selection that has shaped the characteristics of wild species and humans alike (see Mindell 2006; Coyne 2009; Lane 2009; and Fairbanks 2012 for good overviews). Nowhere is the power of natural selection on genetic composition more striking than in the staggering variety of modern dog breeds, and importantly this manifests in not just physical traits but also in behaviour. The same is true of the power of natural selection in shaping the evolutionary history of humans.

Box 2.2: How Deep Is Your Darwinism? (Adapted from Richards 2000)

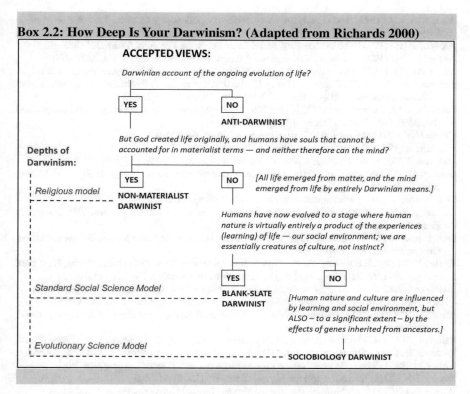

We will explore probable reasons for the contemporary ignorance of—and resistance to—Darwinism later in the book, but it is important to acknowledge at the outset that there are various 'depths of Darwinism' to which both professionals and the general public ascribe (Box 2.2) and that the premise of this book is built upon the deepest: *Sociobiology Darwinism*. This has its roots in seminal accounts by Dobzhansky (1961, 1962, 1963), Alexander (1971, 1974, 1975, 1979), and Wilson (1975, 1978) and is represented by the modern field of evolutionary psychology. According to this interpretation, human minds—ways of thinking and behaving (both characteristically and in their variety)—are products of interacting effects of exposure to particular environments, including opportunity for social learning within those environments, together with effects of genetic inheritance, i.e. the expression of genes that were propelled into future generations as a consequence of Darwinian selection in the ancestral past.

References

Agren JA (2021) An idea with bite: the 'selfish gene' persists for the reason all good scientific metaphors do – it remains a sharp tool for clear thinking. Aeon, 2 Sept 2021, https://aeon.co/essays/why-the-selfish-genes-metaphor-remains-a-powerful-thinking-tool

Alexander RD (1971) The search for an evolutionary philosophy of man. Proc R Soc Victoria 84:99–120

Alexander RD (1974) The evolution of social behavior. Annu Rev Ecol Syst 5:325–383

Alexander RD (1975) The search for a general theory of behaviour. Behav Sci 20:77–100

Alexander RD (1979) Darwinism and human affairs. University of Washington Press, Seattle

Antonovics J, Abbate JL, Baker CH, Daley D, Hood ME et al (2007) Evolution by any other name: antibiotic resistance and avoidance of the E-word. PLoS Biol 5(2):e30

Brinkworth M, Weinert F (eds) (2012) Evolution 2.0: implications of Darwinism in philosophy and the social and natural sciences. Springer, Berlin

Coyne JA (2009) Why evolution is true. Penguin Group, New York

Dawkins R (1976) The selfish gene. Oxford University Press, Oxford

Dawkins R (2004) The ancestor's tale. Houghton Mifflin, New York

Dennett DC (1995) Darwin's dangerous idea: evolution and the meanings of life. Touchstone, New York

Dobzhansky T (1961) Man and natural selection. Am Sci 49:285–299

Dobzhansky T (1962) Mankind evolving: the evolution of the human species. Yale University Press, New Haven

Dobzhansky T (1963) Anthropology and the natural sciences: the problem of human evolution. Curr Anthropol 4: 138+146–148

Fairbanks DJ (2012) Evolving: the human effect and why it matters. Prometheus Books, New York

Lane N (2009) Life ascending: the ten great inventions of evolution. Norton, London

Le Page M (2008) Evolution: 24 myths and misconceptions. New Sci 16 Apr 2008, http://www.new-scientist.com/article/dn13620-evolution-24-myths-and-misconceptions.html#.VONKEfnF_iY

Mindell DP (2006) The evolving world: evolution in everyday life. Harvard University Press, Cambridge

Palumbi SR (2001) Humans as the world's greatest evolutionary force. Science 293:1786–1790

Richards JR (2000) Human nature after Darwin: a philosophical introduction. Routledge, London

Ridley M (2003) The agile gene: how nature turns on nurture. Harper Collins, New York

Stamos DN (2008) Evolution and the big questions. Blackwell, Malden

Wilson EO (1975) Sociobiology: the new synthesis. Harvard University Press, Cambridge

Wilson EO (1978) On human nature. Harvard University Press, Cambridge

Chapter 3
Becoming Human

We are descendants of a long lineage, only a fraction of which is human. (Gray 2002)

Cradle of Mankind (at Olduvai Gorge and Archaeological Museum, Ngorongoro, Tanzania). (Image Courtesy of Alamy (https://www.alamy.com/))

L. Aarssen, *What We Are: The Evolutionary Roots of Our Future*,
https://doi.org/10.1007/978-3-031-05879-0_3

Our species belongs to the family Hominidae—the Great Apes. There are eight of us: three species of orangutans, two species of gorilla, two species of chimpanzee, and humans—*Homo sapiens*. Our genetically nearest relatives are the chimpanzees, with which we shared a common ancestor about 7 million years ago. Many other species of Hominidae, and countless unknown and unnamed intermediates, have come and gone since. Several of these were, like us, 'human' (Box 3.1), which is commonly distinguished (including here) by having an upright bipedal posture—with its earliest appearance in the fossil record from Africa about 6 million years ago.[1] Many additional human traits of course followed, but a few very significant ones explored in Chap. 4—awareness of time, theory of mind, complex language, and culture—apparently ended up earlier and well developed only in our species (Aarssen 2017), the most recently evolved of all humans, appearing in Africa only about 250–350 thousand years ago. The profound fitness advantage afforded by this assemblage of cognitive traits plausibly accounts for why we are the only human species left on the planet, why our ancestral lineage is probably to a large extent responsible for the earlier extinction of other humans (Longrich 2019) who at one time coexisted with us,[2] and why our recent activities are causing the extinction of thousands of other species every year. The other seven Great Apes don't know that their days are probably numbered as well, if modern civilization continues on its present course (Estrada et al. 2017).

---

[1] Some writers refer to bipedal genera older than Homo as 'prehumans' or 'proto-humans'.

[2] Note that there is a long history of debate among paleoanthropologists, referred to as 'lumpers' versus 'splitters' (see commentary by Schwartz 2015). The 'lumpers' promote a linear conception of evolution, a gradual transformation of a limited number of chronologically successive species within the *Homo* genus, or even a view that all fossil specimens of *Homo* should be collapsed into a single species—*Homo sapiens*—with a very wide degree of variability in time and space. The 'splitters', in contrast, recognize greater taxonomic diversity, with several separate species of *Homo* overlapping in time (as depicted in Box <InternalRef RefID="FPar1">3.1).

Box 3.1: Human Evolution. © NR Longrich (from Longrich 2019). Used with Permission

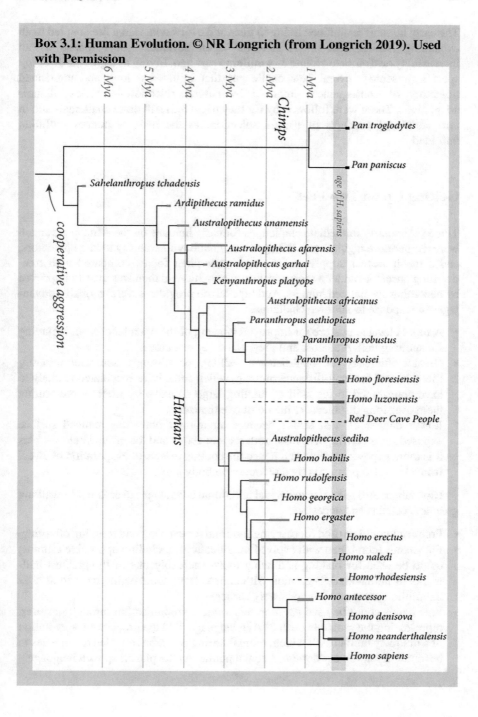

The early humans abandoned living in trees and evolved instead a life adapted to an expanding habitat/niche in Africa—the open savannah—in response to an increasingly arid climate around 3–4 million years ago. Living as a ground-dwelling species presented several new challenges that initiated a long and unplanned trajectory of consequences imposed by natural selection—a series of new adaptations. These were followed by (or they triggered) still more challenges and in turn additional products of natural selection, as the path of human evolution unfolded.

Getting Up on Two Feet

Three particularly immediate challenges for early humans on the African savannah were temperature regulation in the open sun, greater vulnerability to large predators, and a much scarcer supply of fruit and other plant foods, compared with tree-dwelling ancestors under a more mesic climate. Survival then was greatly improved by becoming specialized hunters. And bipedalism provided several probable advantages in response to all three challenges:

- A more elevated posture for improved viewing of the open landscape, including for spotting both predators and prey, and other resources.
- Greater efficiency for long-distance walking, permitting a nomadic/migratory lifestyle, and for long-distance running, which some have speculated is likely to have contributed to the skill of hunting large animals, by literally outrunning them over long distances, to the point of exhaustion.
- Lower danger of heat stroke because an upright body has reduced surface exposed for receiving heat from both the sun above and the ground below—plus it is more exposed to wind and hence the cooling effects of evaporation of sweat from skin (also promoted by a reduction in body hair).

Bipedalism also gave rise to several additional advantages over knuckle-walking great ape cousins and ancestors:

- Free arms could be used for carrying food and for carrying and rescuing offspring.
- Free hands (combined with evolved modifications, including opposable thumbs) could be used for making and using tools—arguably one of the greatest hallmarks of humankind, beginning with simple stone tools dating back to at least 2.5 million years ago (and probably earlier).
- Free hands could be used for more elaborate communication through gestures, representing the probable birth of sign language, and the precursor for evolution of a uniquely human skill which, more than any other, has enabled our species to become the supremely dominant social animal on the planet: spoken language.

Big Brains and Social Intelligence

Increased meat consumption provided an important secondary advantage: large intestinal tracts became less necessary, allowing more nutrition to be allocated to larger brains (Aiello and Wheeler 1995). Mammals with larger brains relative to body size generally have greater behavioural, problem-solving flexibility, with potential fitness benefits to individuals facing novel or altered environmental conditions—the 'brain size-environmental change' hypothesis (Sol et al. 2008). This would have been a crucial adaptation especially as humans (*Homo erectus*) began leaving Africa about 1 million years ago, expanding into Eurasia and encountering a wide range of new landscapes, climates, potential prey species, edible plant species, predators, and other mortality risks. It would also have been a major advantage for our own species as it apparently faced near extinction some 130,000 years ago when a major ice-age drought swept across Africa (Marean 2010) and possibly again about 75,000 years ago during the global cooling episode resulting from the massive Toba volcano eruption on what is now the Indonesian island of Sumatra (Ambrose 1998).

Box 3.2: Human Evolution Based on Skull Endocasts of Fossil Archaic Primates and Early Hominids. Adapted from van Ginneken et al. (2017). Used with Permission

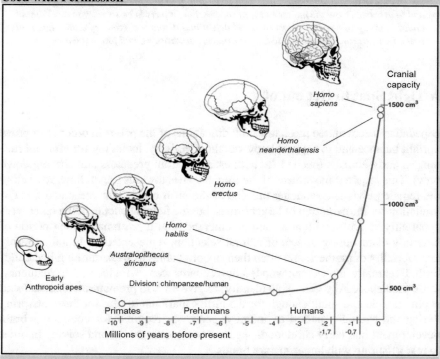

Larger brains also allowed humans to become a social animal unlike any other—the 'brain size-social intelligence' hypothesis (Dunbar 1998). In species of non-human primates, there is a strong correlation between the size of the social group and the relative size of the neocortex—the outer surface layer of the brain primarily responsible for social intelligence (Dunbar 2010). According to this hypothesis, the evolution of larger brains (Box 3.2) was driven to a large extent by the need to live in larger groups, initially at least for protection against large predators. This required a cognitive capacity to process and evaluate constantly changing information about the local social domain (discussed in more detail in Chap. 4). This capacity continued to be vitally important, on an even larger scale, as humans discovered fire and began a campsite-based social life characterized by division of labour and sharing of food and childcare, and also essential for the group cooperation needed to hunt large game, and to attack and defend against rival tribes and clans (Stix 2014). Wilson (2012, p. 44) elaborates:

> Carnivores at campsites are forced to behave in ways not needed by wanderers in the field. They must divide labor: some forage and hunt, others guard the campsite and young. They must share food, both vegetable and animal, in ways that are acceptable to all. Otherwise the bonds that bind them will weaken. Further, the group members inevitably compete with one another, for status or a larger share of food, for access to an available mate, and for a comfortable sleeping place. All of these pressures confer an advantage on those able to read the intention of others, grow in the ability to gain trust and alliance, and manage rivals. Social intelligence was therefore always at a high premium. A sharp sense of empathy can make a huge difference, and with it an ability to manipulate, to gain cooperation and to deceive. To put the matter as simply as possible, it pays to be socially smart. Without doubt, a group of smart prehumans could defeat and displace a group of dumb, ignorant prehumans, as true then as it is today for armies, corporations, and football teams.

A Tight Spot to Get out of

Bipedalism necessitated mechanical modifications of the pelvis in order to improve upright balance and to support body weight effectively, including for efficient running in early humans (needed for both escaping from predators and chasing down prey). This required movement of the ischial spines towards the midline, restricting the female pelvis and reducing the size of the birth canal. This presented a major limitation for the evolution of larger brains, but the fitness benefits of the latter were apparently so profound that a solution emerged after a presumably long period of intensely disfavouring effects of natural selection, represented by brutal suffering and mortality of mothers, and often their unborn babies as well, during failed childbirth. Eventually, the women who left descendants were, with increasing frequency, those predisposed (by genotypic identity) to give birth to premature offspring (with skulls flexible and small enough for the birth canal) and to care for those offspring during the extended period of infant dependency required for continuing brain development into early childhood—again a cost worth paying and solving in order to rear offspring with larger mature brains.

This development had huge implications for subsequent social evolution—favouring, through natural selection, prepared learning and behaviours associated with greater male involvement in parental care, and with more infant care assistance from extended family, as well as from non-kin within the local social group (Hrdy 2009). Some have also suggested that with a longer childhood immaturity period, environmental (epigenetic) variation had greater opportunity to affect the schedules of turning genes on and off while the brain is still developing, thus producing wider variation in adult behaviours that are not simply hardwired by genotypes and in turn allowing expression of more flexible contingent phenotypes within rapidly changing environments (Ellis and Bjorklund 2005).

Firing Up Human Evolution

No other animal has ever controlled and used fire. Its importance in human evolution is paramount (Burton 2009), and larger brains in particular depended on it. Precisely when controlled fire use first began is unknown, but archaeological evidence points to human cooking with fire, probably by *Homo erectus*, at least as early as approximately 1 million years ago (Berna et al. 2012). Cooking meant less time and energy needed for chewing. It increased the digestibility of food, especially for meat protein, and also made meat safer from pathogens and less prone to spoilage. Greater energy available from food, therefore, meant more nutrition that could be used to support a larger brain (Wrangham 2010)—and interestingly, the above discovery, dated at around 1 million years ago, coincides with the approximate timing of a huge jump in human brain size (Box 3.2).

Fire could also be used to ward off predators, and it provided light at night (effectively extending the length of daytime activities), as well as night-time warmth—hence possibly facilitating less need for body hair (thus promoting longer-distance running skill, for chasing prey to exhaustion, without over-heating). Campsites probably also provided a focal point for socializing—gathering people together and fostering a sense of safety and belonging to a community. This in turn may have presented opportunity to reinforce social learning and communication skills, including through story-telling—at first through hand gesturing and other symbolisms and eventually by discovering and practicing effective vocalizations. Congregation and activity around campsites would also have provided opportunity for teaching the value of group identity, cooperation, sharing, assistance, and provisioning for each other in times of scarcity and illness, with a resulting reduction in mortality rates, especially for infants and children.

The Maze of Human Evolution

It is important to emphasize that natural selection has no purpose, and so there was nothing planned or inevitable about the journey of human evolution. As for all species, it has been like a journey through a maze (Box 3.3). Wilson (2012) provides an apt narrative:

Box 3.3: Human Evolution as a Journey Through a Maze

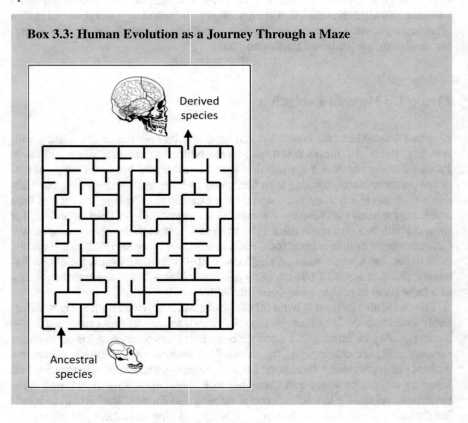

The possible evolution of a species can be visualized as a journey through a maze. As a major advance such as the origin of sociality is approached, each genetic change, each turn in the maze either makes the attainment of that level less likely, or even impossible, or else keeps it open for access to the next turn. In the earliest steps that keep other options alive, there is still a long way to go, and the ultimate far distant attainment is least probable. In the last few steps there is only a short distance to go, and the attainment becomes more probable. The maze itself is subject to evolution along the way. Old corridors, (ecological niches) may close, while new ones may open. The structure of the maze depends in part on who is travelling through it, including each of the species.

In every game of evolutionary chance, played from one generation to the next, a very large number of individuals must live and die. The number however, is not countless. A rough estimate can be made of it, providing at least a plausible order of magnitude guess. For the entire course of evolution, leading from our primitive mammalian forebears of a hundred million years ago to the single lineage that threaded its way to become the first Homo sapiens, the total number of individuals it required might have been one hundred billion. Unknowingly, they all lived and died for us.

Many of the players, among the other evolving species, each containing on average a few thousand breeding individuals per generation, also frequently declined and disappeared. Had this happened to any one of the long line of ancestors leading to Homo sapiens, the human epic would have promptly ended. Our prehuman ancestors were not chosen, nor were they great. They were just lucky.

References

Aarssen LW (2017) The sapiens advantage. Ideas Ecol Evol 10:6–11

Aiello LC, Wheeler P (1995) The expensive-tissue hypothesis: the brain and the digestive system in human and primate evolution. Curr Anthropol 36:199–221

Ambrose SH (1998) Late Pleistocene human population bottlenecks, volcanic winter, and differentiation of modern humans. J Hum Evol 34:623–651. https://doi.org/10.1006/jhev.1998.0219

Berna F, Goldberg P, Kolska Horwitzc L, Brink J, Holt S, Bamford M, Chazang M (2012) Microstratigraphic evidence of in situ fire in the Acheulean strata of Wonderwerk Cave, Northern Cape province, South Africa. Proc Natl Acad Sci 109:E1215–E1220. https://doi.org/10.1073/pnas.1117620109

Burton FD (2009) Fire: the spark that ignited human evolution. University of New Mexico Press, Albuquerque

Dunbar R (1998) The social brain hypothesis. Evol Anthropol 6:178–190

Dunbar R (2010) How many friends does one person need? Dunbar's number and other evolutionary quirks. Faber and Faber, London

Ellis BJ, Bjorklund BF (eds) (2005) Origins of the social mind: evolutionary psychology and child development. The Guilford Press, New York

Estrada A et al (2017) Impending extinction crisis of the world's primates: why primates matter. Sci Adv 3(1):e1600946. https://doi.org/10.1126/sciadv.1600946

Gray J (2002) Straw dogs: thoughts on humans and other animals. Granta, London

Hrdy SB (2009) Mothers and others: the evolutionary origins of mutual understanding. Belknap Press, Cambridge

Longrich NR (2019) Were other humans the first victims of the sixth mass extinction? The Conversation. https://theconversation.com/were-other-humans-the-first-victims-of-the-sixth-mass-extinction-126638

Marean CW (2010) When the sea saved humanity. Sci Am 303:54–61

Schwartz JH (2015) When fiction becomes fact. Q Rev Biol 90:193–198

Sol D, Bacher S, Reader SM, Lefebvre L (2008) Brain size predicts the success of mammal species introduced into novel environments. Am Nat 172:S673–SS71

Stix G (2014) The it factor. Sci Am 311:72–79, 19 Aug 2014. https://doi.org/10.1038/scientificamerican0914-72

van Ginneken V, van Meerveld A, Wijgerde T, Verheij E, de Vries E et al (2017) Hunter-prey correlation between migration routes of African buffaloes and early hominids: evidence for the "out of Africa" hypothesis. Integr Mol Med 4. https://doi.org/10.15761/IMM.1000287

Wilson EO (2012) The social conquest of earth. Liveright Publishing Corporation, New York

Wrangham R (2010) Catching fire: how cooking made us human. Basic Books, New York

Chapter 4
Discovery of Self

MIND, n. A mysterious form of matter secreted by the brain. Its chief activity consists in the endeavor to ascertain its own nature, the futility of the attempt being due to the fact that it has nothing but itself to know itself with. (Bierce 1906)

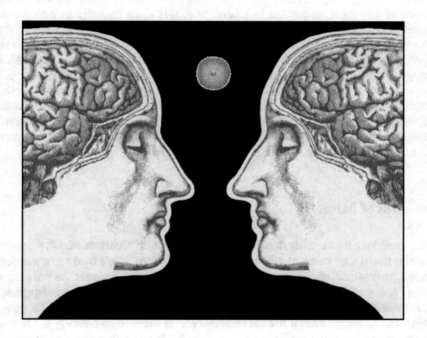

Image reprinted from: S. Laureys (2006) *The Boundaries of Consciousness: Neurobiology and Neuropathology*. Elsevier, Copyright (2022). Used with permission

25
L. Aarssen, *What We Are: The Evolutionary Roots of Our Future*,
https://doi.org/10.1007/978-3-031-05879-0_4

By 100,000 years ago, humans, and *Homo sapiens*, in particular, had evolved cognitive capacities – for intelligence, curiosity, intuition, imagination, computation, logic, reasoning, and memory – and social lives far more sophisticated than any other animal known to exist before or since. It was not just about larger brains but also a greater number of neurons, their interconnections, and where they are located in the brain (Herculano-Houzel 2019).

But mental experience for our species, and possibly ours alone, soon pushed ahead still further with a 'great leap forward' (Diamond 1992). Around at least 40,000–50,000 years ago, but probably much earlier (Tomasello 2014), our ancestors started to become equipped for profound enlightenment – they discovered a sense of time and got an impression of personhood. Variously called the 'human spark', the 'mind's big bang', or 'symbolic self-awareness' (Dobzhansky 1964, Sedikides and Skowronski 1997), individuals began to perceive their identities as 'actors' among many on the 'stage of life' as it passes through time and with recognition of this same perception in others. With this came capacity to plan for the future and to envisage an existence of unseen others and events from the past that was understood to have been just as 'real' as the present. It meant being able to use abstract/symbolic thinking and insight to see beyond the actual to the possible, to imagine what one's predecessors might have thought or done, and to deliberately control one's behaviour with a view extending beyond just the present, while anticipating possible outcomes of one's actions (e.g. resisting temptation in expectation of delayed but larger gratification). Indications of time awareness and individual identity may be found in some other animals but nothing on a scale that compares with the capacity for these cognitive skills in humans.

Mindful of Minds

This great leap forward involves what is referred to as a 'theory of mind': an ability to infer the mental states of others by extrapolating from one's own private experiences, perceptions, and memories. It means being able to interpret the position of others (perspective taking) and so to interpret shared intentionality and also understand that others can and do have intentions that differ from one's own. It means being able to imagine not just what someone else is likely to be thinking and feeling (both now and in the future) but also what she/he is imagining about what *you* are likely to be thinking and feeling (both now and in the future). With this new cognitive toolkit for social intelligence, our species – unlike any other – could predict the behaviour of others and gauge how one is (or might be) perceived by others and hence influence others to perceive us as we wish them to. We could understand sarcasm; manage rivals; build (or avoid) coalitions with certain others; feel a sense of morality, empathy, trust, and friendship towards others; experience emotions such as sympathy, pity, compassion, gratitude, forgiveness, pride, guilt, envy, anger, jealousy, shame, embarrassment, loneliness, and disgust; and recognize these affective states in others. We became ever more skilled experts at cooperation, negotiation, and teamwork and, at the same time, masters of manipulation, deception, cheating, and lying.

Speaking Up

As these social skills were practiced and perfected, a profoundly important advantage was defined and thus strongly favoured by natural selection: ability to communicate with others more directly and more precisely compared with just grunts, screeches, gesturing, and primitive sign language and thus manage our affairs with them more efficiently and effectively. The stage was set for the evolution of spoken language. According to Dunbar (2004), 'Speech and language evolved to enable us to bond social groups that were getting too large to bond by conventional primate grooming'. It was an intellectual/cognitive advance but also anatomical, requiring development and modifications of the larynx, tongue, and associated muscles – perhaps beginning as a musical protolanguage that helped our ancestors avoid predators (Knight and Lewis 2017).

With the capacity for complex language, individuals could quickly plot a strategy together for hunting, brainstorm about how to make a better tool, communicate complex information to children and grandchildren, and quickly learn about whom to trust or emulate from the latest gossip. They could learn from legends and stories about the 'possible' and its consequences and thus become better equipped to respond effectively when the possible became reality. The great survival value of story-telling for our ancestors probably accounts for the universal popularity – today – of entertainments from cultural products like novels, films, and television (Gottschall 2012). Updike (2000) elaborates:

> As to the movies — who of my generation did not seek his innermost self within the glittering, surging, world picture that cinema presented to its rapt receivers in the semidark? What was worthwhile and true was somehow there, coded in Gary Cooper's pale-eyed deadpan and Esther Williams' underwater smile.

These advantages through language allowed humans to master the art of living in ever larger units of social organization – tribes, clans and villages, and eventually cities and states (Gamble et al. 2014). Language also allowed our ancestors to command attention and direct persuasion of large groups for ideological, political, and military goals. It produced a kind of intelligence with impact that went beyond that of a single mind. With our minds, through our language, collectively we have made our world. As Lovelock (2009, p. 256) put it:

> … our amazing achievements come from the additional ability of our brains to communicate and persuade, so that the thoughts of one or a few can persuade the many to lose their identities and act coherently as if they were a single individual. The powerful amplification of the expressed intentions of a tribal leader can always prevail against an incoherent foe or the natural world. This synchronization of will we share with social insects and termites as well as with flocks of birds and fish, and it empowers us far beyond the possibilities of a single isolated intelligence, even one much more able than we are.

A Cultural Thing

The above series of cognitive adaptations, each dependent on preceding products of purposeless natural selection and genetic inheritance, gave rise to perhaps the most astounding advance of all in the history of humanity: a second, entirely new mode of evolution, with a scale and complexity available only to humans – cultural (socially transmitted) inheritance, involving the copying, sharing, and storage of ideas and knowledge (new and old), including local values, beliefs, and customs across (and within) generations, through learning from communication (Boyd and Richerson 2009; Pagel 2012; Laland 2017; Boyd 2018). Interpretations of modern products of culture will be explored in more detail in later chapters, but two of the most conspicuous – religion and art – are evidently tens of thousands of years old. We can reasonably speculate that creative ancestral minds might easily have imagined a sense of 'sacred', involving spiritual worlds and supernatural powers that could be deferred to in accounting for the unknown and unexplained and to which appeals could be made for relief from the hardships of life. And such visionaries might easily have attracted faithful followers. Interpretations of some early products of art, e.g. carved figurines/amulets dating from 25,000–35,000 years ago, point to the possibility of animistic/religious/spiritual symbolism and superstitions. Early religion then probably inspired early art and vice versa (Mithen 1996).

Other forms of early art (dating variously from 10,000 to 50,000 years ago) include cave paintings, etchings and engravings (on stone, shells and bones), stone sculptures, personal ornaments and jewellery made from shells, teeth and ivory, and musical instruments (bone/ivory flutes and rattles) (Dutton 2009; Davies 2012). Some of these and other aesthetic sensibilities and artistic expressions (e.g. body art and piercings, dance, chants, rituals, and ceremonies) may have served in bolstering distinctive identity for group membership, e.g. within a local camp, tribe or clan – i.e. symbols that distinguish 'us' from 'them' (including perhaps as territorial markers). As representations of solidarity and comradery – pride in 'us' – these symbolisms could have been of vital importance for strengthening social cohesion, networking, and coalitions, thus promoting group survival (tribalism), hence individual survival and prosperity and, ultimately, gene transmission success for group members. Personal adornments, like jewellery, may also have been worn as symbols for 'honest' (truthful) advertisement of status or skill (e.g. as a hunter or as a creative mind), and so the evolution of attraction, by the opposite sex, to these and other displays of talent and creativity would have rewarded gene transmission success for the attracted mates, as well as for the attractively adorned and artistically accomplished.

Cultural evolution is distinct (from biological evolution) in that it need not involve genetic change. But *capacity* for human culture and its evolution, are genetically determined; its absence – in its 'degree' – in other species is accounted for only because of missing genes. In addition, as we will see in later chapters, many features of modern human cultures have also been shaped in part by effects of genetic inheritance, as products of natural selection in the ancestral past, but that genetic inheritance, in turn, is also commonly shaped in part by cultural selection (Chap. 6).

The Burden of Discovery

A sense of time and personhood also gave our ancestors what turned out to evoke our most enduring fear – perception of aging and awareness of limits to longevity. 'A being who knows that he will die arose from ancestors who did not know' (Dobzhansky 1967). And once we could foresee our own death, this – through some remarkable coincidence of genetic linkage (still undeciphered) – became tied to a hard-wired 'eventual mortality' anxiety. This is captured whimsically in a definition of life from Bierce (1906): 'LIFE, n. A spiritual pickle preserving the body from decay. We live in daily apprehension of its loss; yet when lost, it is not missed'. Jean-Jacques Rousseau (1754) contrasts this with the inferred primitive human condition:

> ... savage man, subject to few passions, and self-sufficient, had only the sentiments and enlightenment appropriate to that state; he felt only his true needs, took notice of only what he believed he had an interest in seeing; ... The only goods he knows in the universe are food, a woman and rest; the only evils he fears are pain and hunger. ... His soul, agitated by nothing is given over to the single feeling of his own present existence, without any idea of the future, however near it may be, and his projects, as limited as his views, hardly extend to the end of the day.

Some extinct *Homo* species may have known that their lives would one day end but possibly not as early as did *Homo sapiens*. And for those that may have known, there is no reason to be certain that they would have felt any foreboding as a consequence of knowing (Aarssen 2017). They would certainly have had the kind of primal fear that we also have (and share widely with other animals), associated with 'survival drive', i.e. an instinctual anxiety response to an *impending* risk (e.g. hunger/starvation, imminent attack by a rival or predator), thus triggering impulsive action (e.g. 'fight, flight, or freeze') that promotes staying alive in the face of such immediate threats. In contrast, awareness that one will eventually die somehow at some unknown time in the future – *even though safe and well fed* – is merely envisaged, with only suppositions about possible details of the experience. And so, not being a looming risk to one's life, it is unclear whether these possible imaginations by early humans would have been likely to trigger the visceral panic response associated with the primal drive to protect corporeal survival. Reproductive success for early humans would have depended less on speculative ruminations about the distant future and more on being acutely cognizant of, and prepared to defend against, the many clear and present dangers that threatened imminent survival (Aarssen 2017).

Fossil evidence of symbolic cognition (dating back at least 50,000 years), represented by intentional burial of the dead (e.g. Rendu et al. 2014), seems to represent an act of honouring the deceased and ceremonial worshipping of 'forefathers' – rituals that might be expected from distant ancestors anticipating, with trepidation, their own inevitable mortality. As Pinker (1997) put it, 'Ancestor worship must be an appealing idea to those who are about to become ancestors'. However, these rituals indicate only mortality salience, not necessarily anxiety. They may have been associated with just a motivation that provided for sanitation/disinfection or just an expectation of (and preparation for) some kind of afterlife assumption – conjured not necessarily out of fear (personal mortality anxiety) but simply as a creative and intriguing myth for satisfying a curious mind about where people happen to go when they stop moving and breathing. For all we know, these burials by primeval humans may have involved joyful ceremonies about being thankful that a beloved family or tribal member, now expired, no longer needed to endure the routinely brutal hardships of daily existence (just trying to get fed and stay alive) – elicited perhaps by a general inclination to wonder (from inevitable shared experiences) whether it might be 'better never to have been' (Benatar 2006).

A Great Leap for 'Theory of Mind'

Early hominids who worried too much about eventualities in the uncertain distant future probably did not become our ancestors – at least not, I suggest, until *Homo sapiens* evolved a distinctly different kind of angst in response to anticipated eventual mortality. It all began with a distant human ancestor carrying a new mutation, perhaps completing a novel genotype (a series/linkage of particular gene/allele combinations) that transformed 'theory of mind' – taking it beyond just the '…abstract representation of one's own attributes and the use of this representation for effective functioning in affective, motivational, and behavioral domains' (Sedikides and Skowronski 1997). This new genotype, I suggest, informed a deeply ingrained (and delusional) sentiment: *One has a mental life – the 'inner self' – that exists separately and apart from material life, and so, unlike the latter, need not be ephemeral.*

Importantly, however, evolution further revised this genotype, giving it an impulsive uncertainty about this sentiment, i.e. an anxiety not primarily about future inevitable death (loss of material life) but more importantly, I suggest, anxiety about possible annihilation of mental life, 'self-impermanence anxiety' (Aarssen 2010). It is a fear that is rooted not in the literal experience of bodily death, i.e. failing to stay alive per se, but rather a fear of not possessing something of the 'inner self' that can transcend inevitable death (Aarssen and Altman 2006). According to Austrian psychologist Alfred Adler (1870–1937), the 'supreme law' of life is that 'the sense of worth of the self shall not be allowed to be diminished' (quoted from Ansbacher 1985). *Homo sapiens*, then, got self-impermanence anxiety not just because of awareness that time brings eventual death but more specifically because of its frightening potential implication (Aarssen 2010): *In bringing eventual death, will time also inevitably annihilate all that I do, and all that I am? Will my thoughts, perceptions, inclinations, ideas, values, beliefs, attitudes, character, esteem, conscience, pride, ego, intellect, personality, hopes, fears, wishes, dreams, intentions, goals, aspirations, knowledge, skills, and virtues all be lost forever?* It is a worry that one's existence does not and will not 'matter' for anything or anyone – that living it is, and will have been, in vain. It results from a uniquely human capability to recognize, and to feel, a fundamental and terrifying truth regarding virtually all individuals who have ever existed: it is as though they never did. 'I came out of and disappear into the infinite anonymity of space and time' (Fingarette 1996).

The instinctual conviction of a separate mental life – existing independently of material life – and the relentless worry about its possible impermanence are, in my view, the most profound feature, and consequence, of human cognitive evolution. The anticipation of perpetuity and integrity for one's mental life (not one's material life) is what defines, I submit, the most rudimentary, cross-cultural sentiment of 'meaning' for one's existence (Aarssen 2010). Accordingly, a mind – the 'inner self' – troubled by its own possible/probable impermanence is a mind worried about a loss of 'meaning'. In Chaps. 9, 10, and 11, I advance propositions regarding how a deeply ingrained susceptibility to this underlying mental anguish has – paradoxically it would seem – played an important role in rewarding the reproductive success of our ancestors and has thus been a driving force in shaping fundamental adaptive motivations and cultural norms of modern humans. More than any other milestone in the evolution of human minds, including those shaped by our primal drives for survival and sex/mating (Buss 2009), adaptations for managing self-impermanence anxiety, I suggest, account for not just the advance of civilization explored in the next chapter but also now its impending collapse.

In a universe whose size is beyond human imagining, where our world floats like a dust mote in the void of night, men have grown inconceivably lonely. We scan the time scale and the mechanisms of life itself for portents and signs of the invisible. As the only thinking mammals on the planet — perhaps the only thinking animals in the entire sidereal universe — the burden of consciousness has grown heavy upon us. We watch the stars but the signs are uncertain. We uncover the bones of the past and seek for our origins. There is a path there but it appears to wander. The vagaries of the road may have a meaning, however; it is thus we torture ourselves. (Eiseley 1957)

References

Aarssen LW (2010) Darwinism and meaning. Biol Theory 5:296–311

Aarssen LW (2017) The sapiens advantage. Ideas Ecol Evol 10:6–11

Aarssen LW, Altman S (2006) Explaining below- replacement fertility and increasing childlessness in wealthy countries: legacy drive and the "transmission competition" hypothesis. Evol Psychol 4:290–302

Ansbacher HL (1985) The significance of Alfred Adler for the concept of narcissism. Am J Psychiat 142:203–207

Benatar D (2006) Better never to have been: the harm of coming into existence. Oxford University Press, New York

Bierce A (1906) The Cynic's word book. Arthur F. Bird, London. Also published as The Devil's Dictionary. http://www.thedevilsdictionary.com

Boyd B (2018) A different kind of animal: how culture transformed our species. Princeton University Press, Princeton

Boyd B, Richerson PJ (2009) Culture and the evolution of human cooperation. Phil Trans R Soc B 364:3281–3288

Buss D (2009) The great struggles of life: Darwin and the emergence of evolutionary psychology. Am Psychol 64:140–148

Davies S (2012) The artful species: aesthetics, art, and evolution. Oxford University Press, Oxford

Diamond J (1992) The third chimpanzee: the evolution and future of the human animal. HarperCollins, New York

Dobzhansky T (1964) Heredity and the nature of man. New American Library, New York

Dobzhansky T (1967) The biology of ultimate concern. The New American Library, New York

Dunbar R (2004) The human story: a new history of mankind's evolution. Faber and Faber, London

Dutton D (2009) The art instinct. Oxford University Press, Oxford

Eiseley L (1957) The immense journey. Random House, New York

Fingarette H (1996) Death: philosophical soundings. Open Court Publishing, Peru, Illinois

Gamble C, Gowlett J, Dunbar R (2014) Thinking big: how the evolution of social life shaped the human mind. Thames and Hudson, London

Gottschall J (2012) The storytelling animal: how stories make us human. Houghton Mifflin, New York

Herculano-Houzel S (2019) Your big brain makes you human – count your neurons when you count your blessings. The Conversation, 26 Nov 2019. https://theconversation.com/your-big-brain-makes-you-human-count-your-neurons-when-you-count-your-blessings-127398

Knight C, Lewis J (2017) Wild voices: mimicry, reversal, metaphor, and the emergence of language. Curr Anthropol 58:435–453

Laland KN (2017) Darwin's unfinished symphony: how culture made the human mind. Princeton University Press, Princeton

Lovelock J (2009) The vanishing face of Gaia: a final warning. Basic Books, New York

Mithen S (1996) The prehistory of the mind: a search for the origins of art, religion and science. Thames and Hudson, London

Pagel M (2012) Wired for culture: origins of the human social mind. Norton, New York

Pinker S (1997) How the mind works. Norton, New York

Rendu W et al (2014) Evidence supporting an intentional Neandertal burial at La Chapelle-aux-Saints. Proc Natl Acad Sci 111:81–86

Rousseau J-J (1754) Discourse on the origins and foundations of inequality among men. In: Cress DA (ed) (1987) The basic political writings. Hackett Publishing, Indianapolis

Sedikides C, Skowronski JJ (1997) The symbolic self in evolutionary context. Pers Soc Psychol Rev 1:80–102

Tomasello M (2014) A natural history of human thinking. Harvard University Press, Cambridge

Updike J (2000) The tried and the treowe. In: Due considerations: essays and criticism. (2007). Randon House, New York

Chapter 5
The March of Progress

For all our intellect and accomplishment, we cannot seem to make a culture that does not kill almost everything in its wake in the name of growth and progress. (Burr 2008)

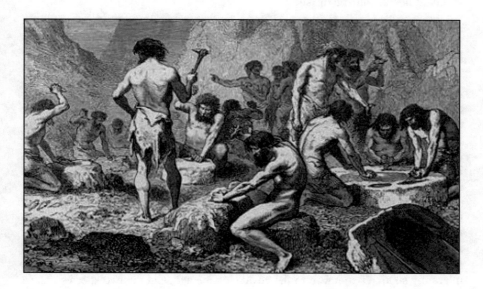

Making flints. (Image courtesy of Getty Images (https://www.gettyimages.ca/detail/news-photo/during-the-stone-age-a-group-of-men-hammer-at-rock-in-order-news-photo/50807369))

In Chap. 4, we saw how the discovery of self and its consequences during the 'great leap forward' forged an epic path of human evolution that gave us culture. This in turn gave birth, over the past 40,000 years or so, to the second great hallmark of humanity, the project of civilization, which continues to this day on its relentless mission of ever greater feats of progress. This chapter is a brief survey of the major milestones in that project.

L. Aarssen, *What We Are: The Evolutionary Roots of Our Future*,
https://doi.org/10.1007/978-3-031-05879-0_5

Handy Tools

For over two million years, tool development had remained largely stagnated; only simple hand tools made of stone were in use. But starting about 300,000 years ago (Tollefson 2018), and especially in the upper Palaeolithic (about 40,000–50,000 years ago), more specialized tools appeared, beginning with more elaborate shapes (e.g. hand axes) and flaked stone with thinner blades, plus tools made from a wider variety of materials, including antlers, ivory, and bone. This was followed by compound tools such as stone axe heads and spear points set in wooden shafts and other weapons for killing animals (and humans) at a distance. These advances led soon to fish hooks, needles, awls, mortars and pestles, braided rope, sewn clothing, and constructed shelters and dwellings and, more recently, computers, satellites, and intercontinental ballistic missiles.

Meeting the Neighbours

Recall from earlier chapters that fitness (gene transmission success) depends directly (for all species) on individual reproductive success, which in turn depends directly on individual survival and prosperity but that the latter, for humans in particular, has in turn been promoted (frequently to a large extent) by the protective and cooperative benefits of belonging to an exclusive social group. As Wilson (2012) put it, 'In its power and universality, the tendency to form groups and then to favour in-group members has the earmarks of instinct'.

This was put to a dramatic test around 35,000–40,000 years ago – but possibly beginning earlier (Harvati et al. 2019) – when *Homo sapiens* extended its range northward into Europe, probably partly in response to severe limitations from a dry African climate at the time. Our ancestors then met their generic cousins – *Homo neanderthalensis* – who had dominated Europe for some time. Neanderthals were migrating south, escaping from the advancing ice age, and the two species met in the Middle East and on the Iberian Peninsula, where, because of lower sea levels, present-day Spain and Morocco were either joined by an isthmus or separated by a very narrow and easily traversed channel. The journey from Africa may have been less perilous for *Homo sapiens*, and so their numbers may have been greater.

By about 28,000 years ago, Neanderthals had gone extinct. The reason(s) remain uncertain – possibly associated with cumulative effects of small population sizes and associated inbreeding (Vaesen et al. 2019). But according to a popular hypothesis, the sapien invaders were at least partially responsible – from a combination of murder, competition for resources and territory, and possibly the introduction of disease epidemics (Diamond 1992). The physical strength and brawn of Neanderthals (which requires a lot of food to support) were possibly no match for the slimmer but smarter (Kochiyama et al. 2018) *Homo sapiens*, who were more advanced in the use of tools – including for fighting, hunting, and harvesting other contested resources. Anatomy may also have played a role; with shorter limbs and a stockier frame, Neanderthals probably had slower running speed, giving them a disadvantage in battle (Steudel-Numbers and Tilkens 2004).

With less-advanced cultural evolution, Neanderthals may have had at least two additional disadvantages. First, they may have practiced little or no division of labour, whereas sapiens are likely to have had only men involved in the dangerous work of hunting large game and defending against rival clans – with women and children engaged in the safer activities of foraging, gathering, and campsite organization. As Walter (2013) put it, 'If men and women both undertook the deadly work of bringing down big game, then women who were killed in hunts would not survive to bear more children. *Homo sapiens* women then would have literally outbred Neanderthal women, while the latter struggled to keep pace with replacing the members they were losing'.

Second, because Neanderthals apparently lacked the tools (e.g. lightweight projectile spears) needed for killing large animals at a distance, they could only strike at close range and were more likely therefore to be injured or killed when hunting. Mortality risks for Neanderthals are also likely to have been generally higher, in part, because of living in a harsher climate affected by the impact of ice ages in Europe. These effects could account for some evidence indicating that Neanderthals generally had a shorter average lifespan (about 30–40 years) (Walter 2013). If true, then natural selection may have favoured faster childhood development, to reach adult size and reproductive maturity earlier, thus increasing gene transmission success by increasing the probability that a son or daughter would, before dying, produce grandchildren and provide adequately for them (including as hunters). However, the shorter childhood for Neanderthals, compared with sapiens (if true), possibly meant that they had less time for engaging in creative childhood play and curiosity and hence less opportunity for developing certain learning skills and social skills prior to meeting the challenges of adulthood and thus less capacity for problem-solving ingenuity. Less time for childhood experiences could mean less variable childhood experiences and hence less opportunity to form unique personalities, resulting in a lower diversity of individual talents within a local population (Walter 2013).

A longer average lifespan for sapiens – possibly to about 60 years – would also have meant that grandparenting was a more significant part of their culture. Importantly, this would have meant greater capacity for accumulation and transmission of knowledge across generations – including important information and skills needed for survival in desperate times (Gopnik 2020). 'Grand-mothering', in particular, also provided supplemental maternal care for grand-offspring. For an older woman, rather than continuing to produce offspring, 'grand-mothering' may in fact have provided a more effective way for promoting the propagation of her gene copies into future generations, thus possibly accounting for the evolution of menopause in human females (Sterelny 2012). In addition, by allowing a personal experience of descendants across not just one, but at least two generations, grand-parenting possibly set the stage for evolution of a fundamentally adaptive motivational domain in modern humans: 'Legacy Drive' (see Chap. 10).

Recent genome sequencing data indicate that there was at least some interbreeding between modern humans and Neanderthals. And so, while the latter are now extinct, some of their genes/alleles nevertheless had transmission (fitness) success, with copies still residing in our species today. Of course this evidence also suggests that Neanderthals were not really a separate species at all.

The Birth of Racial and Cultural Diversity

Cultural and technological advances during the 'great leap forward' allowed modern humans to expand their range across Eurasia and beyond. By about 10,000 years ago, *Homo sapiens* occupied every continent except Antarctica. Along this journey our ancestors encountered wider ranges of living environments, including variation in climate, diet options, and disease-causing pathogens. The landscape was a patchwork of habitat quality; populations thus settled and grew in hospitable ones while remaining more or less isolated for long periods of time from certain other centres of population growth. Different genes conferred local adaptations to the challenges of survival in these different environments, and so – because of geographic isolation – humans started to display heritable between-population variation in morphological and physiological traits, such as body size and shape, skin colour, ability to digest certain foods, resistance to certain diseases, and physiological responses to oxygen levels at different altitudes.

Other differences between peoples living in different regions developed from cultural evolution. Most people spent their entire lives within a short distance from their birth places and would have no opportunity to learn of the existence of people living in distant lands (Diamond 1992). This geographic isolation resulted in the development of wide variation between regional populations in local ideas, knowledge, values, beliefs, customs, rituals, myths, superstitions, religion, language, and products of art and technology – all representing distinctive iconic symbols of social group identity.

More of 'Us' Versus 'Them'

At some point, opportunities started to increase for occasional long-distance travel. Formerly isolated populations therefore came into occasional contact, marking the beginnings of long-distance trade, involving beneficial exchanges of raw materials (e.g. different varieties of stones, gems, and spices) and cultural products (e.g. tools, ornaments, and other works of art). But relationships between neighbouring populations would have been characterized by a shifting balance between friendly trade and ethnocentric/xenophobic hostility (Diamond 1992) – the latter rooted in a deeply instinctual suspicion of, and uneasiness with, 'them', i.e. people who clearly belong to a different social group, signalled by appearances, language, behaviours, and traditions uncharacteristic of one's own social group ('us'). When ancestors encountered examples of 'them', undoubtedly they often turned out to be hostile – or when they left, residents of 'us' commonly became ill and often severely so (from a pathogen brought with 'them' but to which they had evolved resistance). The illness might have been perceived as a curse cast upon 'us' by 'them', as harbingers of evil. It would have been adaptive, therefore, for these ancestors to evolve a behavioural disposition (i.e. informed by effects of both genetic and cultural inheritance) to be at least wary (if not fearful) of interactions with all variations of 'them', both previously and newly encountered, and to regard them by default with distrust, often escalating into pre-emptive hostility that would have been returned in kind (Bowles 2009).

Accordingly, as in the standoff with Neanderthals, much of the history of human conflict within our own species – from territorial border disputes to horrific genocides – can be traced back to these evolutionary roots. In every episode of human history, chronic xenophobia thus fuelled escalating motivations for more and more development of innovative technology for attacking and defending from 'them', including weapons for mass killing (Cronin 2020). These ambitions in turn inspired many other domains of cultural evolution and hence further cultural differentiation between regions, fuelling still more xenophobia. We return to this topic in Chap. 8.

Taming Nature

By around 10,000 years ago humans had so perfected their hunting skills and weaponry that many species of large game had been driven to extinction in Europe, Asia, Australia, and North America – including mammoths, mastodons, giant deer, bison and armadillos, and several species of horses, camels, and oxen. Effects of ice age climate changes may have also played some role, but humans are at least partially if not largely responsible. Yet they could not possibly have known when or whether any one species was at risk across its distribution range or that it even had a limited distribution range. The activities of any given hunting group covered only a relatively small region, and population declines of game species to extinction took place over thousands of years. Humans, therefore, didn't realize it until it was too late. And they probably didn't realize – at least initially – that the cause had anything to do with too much hunting.

The stage was set then for seeking an alternative to hunting and foraging as a principal way of life. Our ancestors must have gotten their first clue for this while they were still primarily hunter-gatherers, when they made their first alliance with a wild species – the wolf. This gave humans their first glimpse of how to tame nature. The proto-wolf/dog, however, initiated its own domestication, probably without any human forethought, intention, or intervention. Some wolves must have had behavioural dispositions (genetically endowed) that made them relatively tame and curious about human activity around hunting kill sites and around campsites, where food was obviously available in relative abundance. Those individuals that showed up behaving more like a friend than a dangerous predator would have presented as objects of curiosity and entertainment for humans, who in turn would have been inclined to reward the animals with food. This of course led to a more permanent mutualism as dogs evolved unique social skills (not shared with their canine ancestors) for interaction with humans, and as humans came to realize that dogs had value that was far greater than just entertainment and companionship, they were much better equipped than humans for tracking, herding, and hunting game and for warning and guarding against danger, especially at night.

Seeing the results of this relationship with dogs undoubtedly inspired some of our ancestors to imagine how other useful animals might be controlled for purposes of human welfare. It began probably first with nomadic pastoralism. Hunter-gatherers, having learned how to follow animal migrations, began to manage flocks of wild fowl and herds of wild goats, sheep, and cattle by guiding them across the landscape in search of green pastures. At some point, however, people discovered a more efficient alternative: collecting, corralling, storing, growing, and enslaving other species locally – both animals and plants – in order to more precisely control the products of their growth and reproduction for exclusive and more reliable use by humans. And they soon learned that even greater benefits resulted when the most productive individuals and varieties, or their products of reproduction, were preferentially selected for breeding or planting in producing the next generation/crop.

Domestication probably also reinforced a deeper meaning for (and implication of) fatherhood: female animals – like female humans – did not give birth when they were isolated from males, i.e. children are born not just of women, and the birth of a particular child required contribution from a particular male, that results from 'lying with a woman'. We don't know of course when in our ancestral past this 'reproductive consciousness' became widely understood, but when it did, it had profound effects on cultural evolution (Dunsworth and Buchanan 2017). One particularly important consequence is that men started to wonder about which children (if any) came from their contribution. The stage was set therefore for natural selection to favour male behaviours and motivations that helped to reinforce certainty of paternity. As we will see in Chap. 7, this evolved into one of the darker sides of human nature and culture.

The Rise of Agriculturalists

With animal domestication and agriculture, modern civilization was born, and the pace of its progress has been ramping up ever since. The first farmers were in areas of fertile soil with dependable water supply from irrigation on relatively level land with moderate rainfall (thus preventing soil from washing away) – Mesopotamia (the fertile crescent), the Nile Valley (ancient Egypt), and the Indus Valley (ancient India) (Burr 2008). Farming and associated cities also developed independently at about the same time in the Far East, Mesoamerica, and South America (Wright 2004).

Farming, including animal domestication, provided a surplus food supply and hence a huge boost to human carrying capacity (potential population size). Because more people could be fed, more could survive to reach reproductive maturity and thus have babies. And because the same land could be farmed for several generations, people could establish larger and more stable communities and eventually cites with permanent houses, governments, and regular channels of trade and commerce. Religion therefore became more organized, and the temple became a centre of institutional power responsible for administrating affairs affecting the whole community (Korten 2006).

Farming also enabled a few to provide food for many, thus allowing some citizens to have specialized roles, devoting time for generating ever greater feats of innovation and accomplishment – in technology, commerce, social services, and politics – and for training to become professional killers, in the military. Some (more privileged) city dwellers would also have had more time than others to engage in production of new cultural products – in art, music, plays, and (eventually) literature.

But several consequences of agriculture presented disadvantages compared with the hunting and foraging way of life (Diamond 1992) Early city dwellers lived in close proximity to each other and hence each other's sewage, and also wastes from domestic animals, with which people also shared close quarters (along with their parasites and infectious diseases). People therefore were constantly re-infected by each other and from effects of poor sanitation. With a large population dependent on relatively few crop species and varieties, agriculture also brought greater risk of large-scale famine due to periodic crop failures. And the farming diet was less varied, concentrating on high carbohydrate foods. Hunter-gatherers in contrast probably had a diet with higher protein and a better balance of nutrients and so were generally healthier and suffered less from illness and disease.

Because agriculture generated surpluses, it became possible for the powerful to expropriate the surpluses for their exclusive use (Korten 2006). It was the dawn of class divisions – the 'haves' (the healthy, wealthy, non-producing elite) and the 'have-nots' (the disease-ridden, poverty-stricken, labouring masses) (Diamond 1992). Kings, religious officials, and other social parasites grew fat on food seized from peasant farmers and tax collected from everyone. To protect the surpluses from outsiders and thieves, they had to be locked up, and so in order to eat and survive, one had to work within the system in order to earn access to the surplus (Burr 2008). This class system never developed in hunter-gatherer societies because virtually every healthy adult was required to participate in gathering or hunting. Some authors (e.g. Diamond 1992) also suggest that agriculture exacerbated sexual inequality, where women often became essentially beasts of burden in crop production and were drained by more frequent pregnancies and childbearing compared with hunter-gatherers.[1]

Agriculture routinely created population overshoot and thus the need to expand boundaries in order to grow more crops in neighbouring lands. Initially this meant in the neighbouring territories of herders and pastoralists, resulting in conflict between them and farmers. Agricultural societies had larger population sizes and hence larger armies, in part because they had larger food surpluses but also because women could have more babies. Nomadic hunter-gatherers, in contrast, kept their children spaced at about 4-year intervals, by extended breast feeding (and hence delayed menstruation) and even infanticide, since mothers must carry children that are too young to walk at the pace of adults. With sedentary farming, however, women could routinely bear a child every 2 years. Hence bands from farming

---

[1] However, according to others (Korten 2006), early agricultural societies (ca 10,000–7000 years ago) were egalitarian and matrilineal (tracing descent through the woman), inspired by recognition that 'life was born of women' – and evidence of goddess worship is common from this time period, where women presumably had leadership roles, including in the temple.

societies out-bred and then drove off or killed bands that chose to remain hunter-gatherers or nomadic pastoralists. Ten farmers, often malnourished, could nevertheless outfight one healthy hunter (Diamond 1992).[2]

An intriguing question is whether agriculture ever really provided more food per mouth in the long run or improved the average quality of life, compared with hunter-gatherers (Box 5.1). Agriculture routinely provided more food but also resulted in more mouths to feed, which inevitably meant a return to hunger, especially when crops failed, as they periodically did. And many other consequences of agriculture (discussed above) also placed severe limits on quality of life. Agriculture certainly created a higher maximum quality of life but also probably a lower minimum quality of life, i.e. with a wider range, and hence perhaps no change in the average (and possibly even a lower median and mode). It is also plausible that hunter-gatherers may have had more leisure time compared with the frenetic pace of life in a modern crowded city.

By around 5000 years ago, however, the agricultural revolution had turned out four important innovations that pushed the project of civilization into a higher gear: metallurgy, domesticated horses, wheeled vehicles, and the written text. With written language, humans, like no other species, could now learn from long-dead ancestors, from thousands of teachers, instead of just from parents and a handful of contemporaries. They could now improve upon past innovations without first having to reinvent them. By the time the first written texts emerged, the foundation events of human culture – the transformation from the hundreds of thousands of years of the tribal life of hunter-gatherers into the agricultural revolution – were long forgotten. As far as the first writers knew, '… humans came into existence farming just as deer came into existence browsing. As they saw it, agriculture and civilization were just as innately human as thought or speech. Our hunter-gatherer past was not just forgotten, it was unimaginable. The Great Forgetting was unknowingly woven into the fabric of the teachings of the foundation thinkers of our culture: Herodotus, Confucius, Abraham, Anaximander, Pythagoras, Socrates and many others' (Burr 2008).

Tools made of bronze and eventually iron and steel brought enormous advantages, especially for agriculture and weaponry and especially when combined with the domesticated horse. The latter together with wheeled vehicles also meant a major boost in capacities for travel and trade. Increased agricultural productivity allowed the feeding of not just larger populations but also larger armies, giving rise to the organized horse-powered military machine of the 'conquering culture' that has dominated the history of civilization ever since – except now with horses replaced by tanks, fighter jets, and nuclear-powered submarines.

---

[2] Some, however (Korten 2006), suggest that instead, the early settled agriculturalists were relatively peaceful and egalitarian and were eventually taken over by the more violent male-dominated nomadic pastoralists who appropriated by conquest the lands and labour of the more prosperous agriculturalists between ca 7000 and 5000 ya. According to this hypothesis, a culture that honoured the power to give life was replaced by a culture that honoured the power to take life – the warrior culture. Some view this takeover (rather than the development of class divisions within the agriculturalists) as the dawn of the era of Empire.

Box 5.1: Did Agriculture Increase the Average Quality of Life?

Agriculture resulted in more food for people, but also more people — hence more crowding. Did this significantly increase the average quality of life? — or only its variance?

Images courtesy of: Canadian Museum of History – Harvesting wheat S97-10796, CD2001-0281-015 – Painting by Winnifred Needler, photo by Harry Foster (https://www.historymuseum.ca/cmc/exhibitions/civil/egypt/egcgeo3e.html); and NASA/JPL-Caltech (https://mars.nasa.gov/gallery/artwork/neanderthals.html)

Empire

The social structure of 'haves' and 'have-nots' in the first cities eventually expanded in scale to create full blown imperialism – the rule of many by a few, based on domination and subordination. Chiefs and eventually emperors, pharaohs, and kings (often self-appointed as both political and religious leaders) provided favours for priests who affirmed, for the general public, the divine nature of the chief's appointment to legitimate the extraction of taxes (Korten 2006). In return for this service, the priests received a share of the tribute as did other supporters, collectors, loyal warriors, warlords, and sheriffs of the chief. With writing, rulers could establish permanent laws, defining precisely what behaviours and activities they wanted to control and punish (Burr 2008). The greater the coercive power of the ruler and his entourage, the greater the temptation to abuse this power for personal self-indulgence (Korten 2006). War, barbarism, and genocide became popular with those subjects who faithfully served the ruler because they received a share of the spoils of conquered peoples. Corruption was inevitable. And it was all made possible because of agriculture.

Rather than killing captives, rulers learned that they could 'domesticate' people – creating serfdom and slavery. 'There is no elite class without a servant class. The maintenance of a dominator system depends on violence or the threat of violence to maintain the extreme class division' (Korten 2006). It involved the reckless squandering of lives and resources and the environment to support the '… privilege and extravagance of the few' (Korten 2006). Societies with greater inequality (between rich and poor) thus generated and maintained masses of deprived people who were eager to strike out in search of new land and opportunities, thereby possibly spreading their culture faster (including its tendency to form class divisions) – not despite the suffering it creates but because of it (Rogers et al. 2011). From Burr (2008):

> Without a standing army, a king is just a windbag in fancy clothes. But with a standing army, a king can impose his will on his enemies and engrave his name in history – and absolutely the only names we have from this era are the names of conquering kings. No scientists, no prophets, just conquerors. For the first time in human history, the important people are the people with armies. After this point, military needs became the chief stimulus for technological advance.

As Charles Galton Darwin (1952) put it: 'Civilization might, loosely speaking, be counted as a sort of domestication, in that it imposes on man conditions not at all typical of wild life'.

Much of this early saga of the conquerors played out in the Middle East and western Europe where cultural evolution proceeded rapidly, particularly because of the circumstances of geography that allowed the advantages associated with the origin and development of agriculture, animal domestication, and metallurgy. Continents differed in the wild plant and animal species that proved useful for domestication (sheep, goats, pigs, cows, and horses in Eurasia by 4000 BC) and in the ease with which domestic species could spread from one area to another (Diamond 1992). Mass extinction of big mammals in the Americas and Australia, probably caused by humans, may have also limited the opportunities for animal domestication there. The most advanced cultures developed the largest groups which through the history of association with animal domestication promoted evolution of infectious diseases and then evolution of resistance to these diseases – providing in turn a further advantage when conquering less advanced cultures comprised of smaller and more scattered groups (because the latter had little or no resistance to infectious diseases) (Diamond 1997).

During the first millennium BC, cultural evolution produced a variety of religions based on salvation – Judaism, Hinduism, Shintoism, and Buddhism. As Burr (2008) notes, 'Earlier gods had been talismanic gods of kitchen and crop, mining and mist, house painting and herding, stroked at need like lucky charms, and earlier religions had been state religions, part of the apparatus of sovereignty and governance, as is apparent from their temples, built for royal ceremonies, not for popular public devotions'. The peak of Empire unfolded over an 800-year reign of Roman control over the Middle East and most of Europe, where it emerged as a champion of Christianity before falling apart by 500 AD after repeated challenges by several barbarian tribes from across Europe and Asia.

Middle Ages

After the fall of Rome, Europe had no individual ruler or nation that was able to establish dominion over the whole. Only the Church of Rome, with its wide reach of influence over spiritual life, provided some unifying force in modulating the competing interests of secular rulers. Power, however, was fragmented among competing factions of Islamic and western Christian empires and Byzantines (a mostly Greek-speaking offshoot of the Roman empire) (Korten 2006). Roman culture and public infrastructure went into decline, and the European continent was thrown into a harsh period of human suffering and misery on a massive scale, lasting almost a 1000 years – from a combination of political unrest, religious wars, famine, starvation, and disease (resulting especially from overcrowding and poor sanitation). The latter was marked by the worst plague in human history, the Black Death, caused by a bacterium transmitted by fleas from rats to humans. It lasted for over 200 years and killed an estimated 75 million people worldwide, reducing the European population by about 50%. There was also a litany of horrors from 'the Inquisition' of the

Roman Catholic Church, which lasted in several forms until the early nineteenth century. As Burr (2008) writes, 'Inquisitors develop a novel technique to combat heresy and witchcraft – torturing suspects until they implicate others, who are tortured until they implicate others, ad infinitum'.

Religions and their bloody conflicts had dominated culture throughout the Middle Ages. It was easy for the ordinary people of empire – the slaves, the conquered, the peasants, the unenfranchised masses – as Burr (2008) put it, '… to envision humankind as innately flawed and to envision themselves as sinners in need of rescue from eternal damnation. They were eager to despise the world and to dream of a blissful afterlife in which the poor and the humble of this world would be exalted over the proud and the powerful. … People everywhere now had salvationist religions to show them how to understand and deal with the inevitable discomfort of being alive'.

But by ca. 1400 AD, the 'city-states' of the Middle Ages, dominated by chaos and competition among feudal lords and religious factions, started to become consolidated within larger and more stable nation states under the power of monarchies (Korten 2006). Cultures started to become less defined by superstitions and blind faith in ancient religions and more by reasoning and enquiring minds. It was the Renaissance, the age of discovery and enlightenment, bringing an explosion of invention – including portable firearms, art, science, and political, social, and religious revolutions. Advanced technology for weapons and large ship building allowed empires to expand worldwide. Transportation of armies of men and horses across oceans placed Europe in an advantageous position to conquer, dominate, and colonize other regions and to expropriate resources from the new world. The grand mission of civilization continued unabated: feeding more of 'us' and fighting more of 'them'.

The Fossil Fuel Party

The discovery of fossil fuels in the 1700s launched a new higher-order scale of hallmarks in human progress, involving massive increases in the rates of advance in technology for agriculture, industry, and urban expansion – triggering unprecedented rates of human population growth. With the discovery of coal, there was less reliance on wood from forests for heating; hence more land could be cleared for agriculture. Because fabrics could be made from oil, there was less need for sheep-grazing land and cotton growing land, thus allowing the conversion of even more non-food cropland to food production. With the commercial production of coal and eventually oil (in the mid-1800s), industry and agriculture went into high gear from fossil fuel energy, giving human carrying capacity another major boost: world population doubled between 1650 and 1850 and doubled again by 1930. But as Burr (2008) writes, this industrial revolution:

… did not bring ease and prosperity to the masses, rather it brought utterly heartless and grasping exploitation, with women and small children working 12 or more hours a day for starvation wages in sweatshops, factories and mines. … As cities became more crowded, human anguish reached highs that would have been unimaginable in previous ages, with hundreds of millions inhabiting slums of inconceivable squalor, prey to disease borne by rats and contaminated water, without education or means of betterment. Hopeless and frustrated, people everywhere became rebellious, and governments everywhere answered with systematic, repression, brutality, and tyranny. General uprisings, peasant uprisings, colonial uprisings, slave uprisings, worker uprisings – there were hundreds … it was the age of revolutions. Tens of millions of people died in them. … People no longer went to war to defend their religious beliefs. They still had them, still clung to them, but … more pressing material concerns had rendered disputes that once seemed so murderously important irrelevant. The hopes that had been invested in religion in former ages were in this age being invested in revolution and political reform. The promise of 'pie in the sky when you die' was no longer enough to make the misery of life in the cauldron endurable.

Larger populations also increased the breeding grounds and evolution of human disease organisms. And the larger acreage of monoculture food crops that supported larger populations increased the breeding grounds and evolution of crop disease organisms. And so, despite technological advances from agriculture and industry, between 1700 and 1900, disease and famine remained rampant. Burr (2008) summarizes some highlights: Deaths by starvation due to crop failures, e.g. ten million in Bengal in 1769; two million in Ireland and Russia in 1845–1846; 15 million in China and India in 1876–1879; and hundreds of thousands at various times in France, Germany, Italy, Britain, Japan, and elsewhere around the world. 60 million Europeans died of smallpox in the eighteenth century, and hundreds of millions died of cholera epidemics, plague, yellow fever, scarlet fever, and influenza around the world.

Over the past 150 years, technology for medical discoveries and treatments advanced in pace with energy technology, resulting in vast increases in the sizes and reproductive potentials of populations, and agricultural technology in turn kept pace with the latter. Death rates decreased and birth rates stayed high or increased. Women with healthier bodies are not only more likely to have a successful pregnancy and delivery, but they also live longer and so they have more years available for pregnancies and deliveries. And healthier children are also more likely to survive to reproductive age. Population explosion was inevitable: world population has more than quadrupled in the seven generations since 1800. Governments, especially in the developed world, welcomed it vigorously; larger populations meant more taxes to collect and larger armies.

A Relentless Reach for Higher Carrying Capacity

As we have seen, the history of civilization can be understood as a story of perpetual escalation in human carrying capacity, i.e. the feeding and protection of more and more people (especially 'us') so that they can live longer and have more and more babies that also survive longer, to do it all over again (Box 5.2). This was accomplished by an unwavering mission for the discovery of any and all sources of energy, especially when our ancestors discovered 'stored sun' (fossil fuels), and the innovation of new and improved ways of capturing it and using it to harvest the resources needed for feeding and protection of members of the social group. These events unfolded in three conspicuous pulses over the course of human evolution (Box 5.3). The result is a kind of never-ending 'arms race': higher carrying capacity alleviates crowding (overpopulation), but with every increase in carrying capacity, population size inevitably rises to meet it, thus requiring still higher carrying capacity to alleviate crowding yet again.

The story of civilization then plays out in some ways like the process of natural selection, i.e. those peoples who engaged with and achieved this escalation more successfully propelled more copies of their genes into future generations, and – because of inevitable competition for limited resources – routinely excluded those peoples who engaged less successfully and hence whose gene copies were thus less successfully propelled or not at all. Also like natural selection, the march of human 'progress' never really ends, and while it commonly leads to greater complexity, it is neither intrinsically 'better' nor permanent. Like the long extinct, Irish elk (*Megaloceros giganteus*), whose antlers, because of 'runaway selection', apparently got too big for its own good, the march of progress, like a runaway train, is going ever faster – too fast and too far for the good of humankind. The grandson of Charles Darwin – Charles Galton Darwin (1952) – wrote almost prophetically over half a century ago:

> Civilization has taught man how to live in dense crowds, and by that very fact those crowds are likely ultimately to constitute a majority of the world's population. Already there are many who prefer this crowded life, but there are others who do not, and these will gradually be eliminated. Life in the crowded conditions of cities has many unattractive features, but in the long run these may be overcome, not so much by altering them, but simply by changing the human race into liking them.

Box 5.2: Some Major Technological Advances in the 'March of Progress' and the Escalation of Human Carrying Capacity

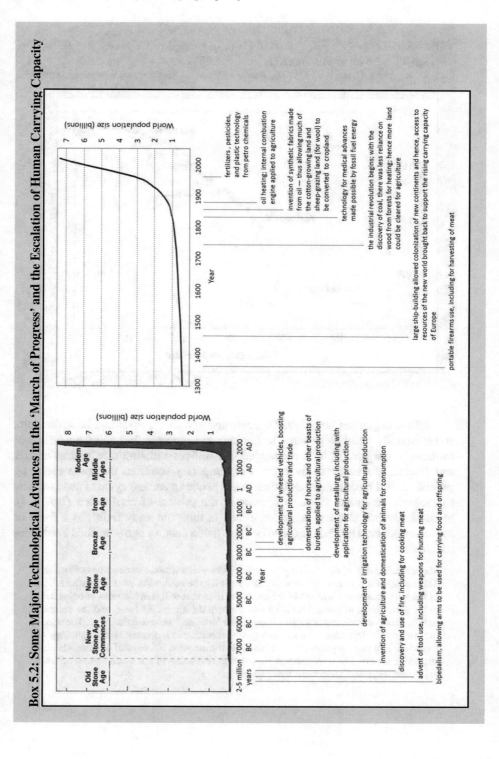

Box 5.3: Three Growth Pulses in Human Carrying Capacity Resulting from Three Major Historical Episodes of Technological Advance
Note log scales. [from Kates (1997); image courtesy of MIT press, Cambridge, MA].

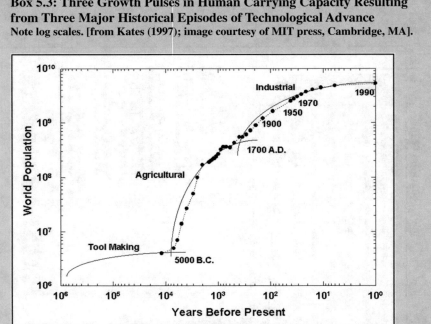

And so, beginning about 10,000 years ago – with the invention of agriculture, and eventually fossil fuel energy technology to support both it and the advance of medicine – our species became trapped in a self-perpetuating race against hunger and disease, characterized by a series of temporary solutions that each ended up generating still more hunger and disease and hence an endless cycle of need for new technologies for new solutions to support the project of civilization (Box 5.4). Medical advance also has an added effect, in this 'run-away train', as a force of natural selection. Dobzhansky (1961) – also half a century ago – reached a sobering conclusion:

Medicine, hygiene, civilized living save many lives which would otherwise be extinguished. This situation is here to stay; we would not want it to be otherwise, even if we could. Some of the lives thus saved will, however, engender lives that will stand in need of being saved in the generations to come. Can it be that we help the ailing, the lame, and the deformed only to make our descendants more ailing, more lame, and more deformed? ... The remedy for our genetic dependence on technology and medicine is more, not less, technology and medicine. You may, if you wish, feel nostalgic for the good old days of our cave-dwelling ancestors; the point of no return was passed in the evolution of our species many millennia before anyone could know what was happening.

Box 5.4: The 'Runaway Train' of Human Civilization
Image source: http://theragblog.blogspot.com/2009/11/barack-obama-stop-runaway-train.html

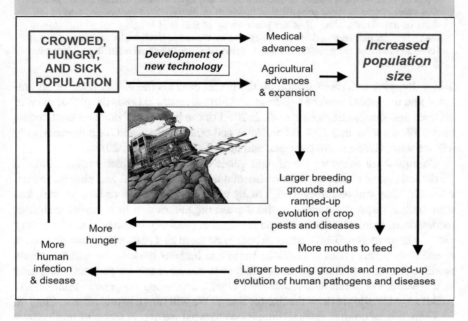

The fossil fuel party, however, is soon coming to an end, and when it does, all indications point to an inevitable crash in human carrying capacity and hence the global human population size. Dobzhansky (1961) ruminates: 'If mankind will prove unable to save itself from being choked by crowding it hardly needs to worry about its genetic quality'. The notion of limits to human population size of course never occurred to *Homo sapiens* over practically all of its evolutionary history. Approaching the end of prehistory (about 5000 years ago), very few of our ancestors would have known with any certainty or even contemplated that there was a world of limited size. And prior to the turn of the last century, a salience of global limits, although known to science, was absent for the vast majority of global citizens preoccupied with meeting the challenges of just being alive.

We make sense of this frenetic march of human progress only by understanding how and why Darwinian evolution shaped us to become what we are. Along this journey, we became a new human subspecies of sorts. We might call it *Homo sapiens totalitarian*. Both our agriculture and our medicine have always been totalitarian (Burr 2008), i.e. they are based on the premise that all other species can be subordinated – used, harvested, killed, eaten, disposed of, or subjected to lethal experiments as needed – for the benefit of human diet, health, and other prosperities. As Burr (2008) put it:

> … our food race is rapidly converting our planet's biomass into human mass. This is what happens when we clear a piece of land of wildlife and replant it with human crops. The land was supporting a biomass comprised of hundreds of thousands of species and tens of millions of individuals. Now all of the productivity of that land is being turned into human mass, literally into human flesh. Everyday all over the world biodiversity is disappearing as more and more of our planet's biomass is being turned into human mass. This is what the food race is about.

Today, the global mass of plastic on the planet exceeds the global mass of all animals, and the global mass of buildings and infrastructure exceeds the global mass of all trees and shrubs (Elhacham et al. 2020). For the biomass of birds on earth today, only 30% are wild, and 70% are chickens and other poultry, and for mammals, only 4% are wild, and 96% are livestock and humans (Bar-On et al. 2018).

Chimps share about 98.4% of their genes with humans, which makes chimps a particularly good choice as an experimental animal for medical and pharmaceutical research. They can be infected with many of the same diseases as humans can, and their bodies respond similarly to the disease organisms and to potential treatment medications. Ironically, for the same reason, i.e. precisely because chimps are very close to humans genetically, many people view them as a particularly bad choice for research use. This raises a moral dilemma that harkens back to the ancient human predilection for 'us versus them' conflict: does one have greater personal obligation to the welfare of some individuals over others, depending on genetic relatedness? And if so, where do we draw the line (or lines)? We will return to this issue in Chap. 8.

There are, of course, several dimensions to what we are, in terms of motivations as well as obligations, including both moral and immoral ones – and the roles of natural selection in shaping these are explored in the next five chapters. But one of the things that we 'are' is particularly relevant as a conclusion to the present chapter, as videographer and social commentator/philosopher, Jason Silva (2014) put it, 'We are addicted to the new'. Walter (2013) elaborates:

> Once the human brain materialized in the form that we now know, outfitted with its genius for creating and shuffling the symbols that make language, imaginary worlds, and above all … the anatomically invisible, terribly murky thing called 'I', creatures emerged that could dream, act on their dreams, and share them with other 'I's around them. And that changed the world.
>
> Our special talent isn't simply that we can conjure symbols, or even weave elaborate, illusory tapestries of them, but that we can share these with one another, roping together both our 'selves' and our imaginings, linking uncounted minds into rambunctious networks where thoughts and insights, feelings and emotions, breed still more ideas to be further shared.
>
> Creativity is contagious this way, and once a light emerged, it must have gone off like fireworks. This has made every human a kind of neuron in a vast brain of humans, jabbering and bristling with creativity, pooling, pulling, and bonding ideas into that elaborate, rambling edifice we have come to call human civilization.

References

Bar-On Y, Phillips R, Miloa R (2018) The biomass distribution on Earth. Proc Natl Acad Sci U S A 115:6506–6511

Bowles S (2009) Did warfare among ancestral hunter-gatherers affect the evolution of human social behaviors? Science 324:1293–1298

Burr C (2008) Culture quake: your children's real future. Trafford Publishing, Bloomington

Cronin AK (2020) Power to the people: how open technological innovation is arming tomorrow's terrorists. Oxford University Press, New York

Darwin CG (1952) The next million years. Doubleday, New York

Diamond J (1992) The third chimpanzee: the evolution and future of the human animal. Harper, New York

Diamond J (1997) Guns, germs, and steel: the fates of human societies. W.W. Norton and Company, New York

Dobzhansky T (1961) Man and natural selection. Am Sci 49:285–299

Dunsworth H, Buchanan A (2017) Sex makes babies: as far as we can tell, no other animal knows this. Did our understanding of baby-making change the course of human history? https://aeon.co/essays/i-think-i-know-where-babies-come-from-therefore-i-am-human

Elhacham E, Ben-Uri L, Grozovski J, Bar-On YM, Milo R (2020) Global human-made mass exceeds all living biomass. Nature 588:442–444

Gopnik A (2020) Vulnerable yet vital: the dance of love and lore between grandparent and grandchild is at the centre, not the fringes, of our evolutionary story. Aeon, 9 Nov 2020. https://aeon.co/essays/why-childhood-and-old-age-are-key-to-our-human-capacities

Harvati K, Röding C, Bosman AM et al (2019) Apidima Cave fossils provide earliest evidence of *Homo sapiens* in Eurasia. Nature 571:500–504. z

Kates RW (1997) Population, technology, and the human environment: a thread through time. In: Ausubel JH, Langford HD (eds) Technological trajectories and the human environment. National Academy Press, Washington, DC, pp 33–55

Kochiyama T, Ogihara N, Tanabe HC et al (2018) Reconstructing the Neanderthal brain using computational anatomy. Sci Rep 8:6296

Korten DC (2006) The great turning: from empire to Earth community. Berrett-Koehler Publishers, San Francisco

Rogers DS, Deshpande O, Feldman MW (2011) The spread of inequality. PLoS One 6(9):e24683. https://doi.org/10.1371/journal.pone.0024683

Silva J (2014) We are addicted to the new. https://www.youtube.com/watch?v=sOlGw9VESiQ

Sterelny K (2012) The evolved apprentice: how evolution made humans unique. MIT Press, Cambridge

Steudel-Numbers KL, Tilkens MJ (2004) The effect of lower limb length on the energetic cost of locomotion: implications for fossil hominins. J Hum Evol 47:95–109

Tollefson J (2018) Advances in human behaviour came surprisingly early in Stone Age. Nature 555:424–425

Vaesen K, Scherjon F, Hemerik L, Verpoorte A (2019) Inbreeding, Allee effects and stochasticity might be sufficient to account for Neanderthal extinction. PLoS One 14(11):e0225117. https://doi.org/10.1371/journal.pone.0225117

Walter C (2013) The last ape standing: the seven-million-year story of how and why we survived. Walker Publishing Company, New York

Wilson EO (2012) The social conquest of Earth. Liveright Publishing Corporation, New York

Wright R (2004) A short history of progress. House of Anansi Press, Toronto

Chapter 6
Whispering Genes

Our genes whisper to us; they do not bark orders. (Barash 2014)

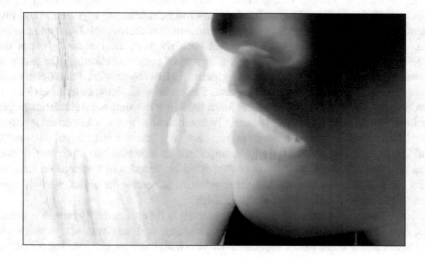

Image source: http://moicani.over-blog.com/article-let-me-whisper-in-your-ear-106830298.html

Throughout our evolutionary history – from the point of becoming human (Chap. 3) to the discovery of 'self' and the emergence of culture (Chap. 4), and all throughout the 'march of progress' (Chap. 5) – Darwinian natural selection has been at work, playing an important role in shaping what we are: our minds, the ways that we think and behave, our cultures, and our civilization.

A mind shaped, even partially, by natural selection, however, has long been and still is an unpopular view among many scholars in the humanities and social sciences (Takacs 2018) and remains today unappreciated (or misunderstood) by the vast majority of the general public. Knowledge of our distinctive capacity for social learning and cultural evolution (Chap. 4), and our ability across the history of civilization to control physical forces of nature and dominate virtually all other life forms on the planet (Chap. 5), have shaped a long and still vibrant tradition of human exceptionalism and socio-cognitive agency, framed by a firm belief in *tabula rasa* – the view that the human mind is a 'blank slate' at birth (see Box 2.2). According to this 'standard social science model', our understanding of the world, and the manner in which we think and behave, is acquired only through exposure to and learning from our environment, including especially the social environment. Variations in human behaviours then are considered to be entirely a consequence of variable environments and variation in the opportunity to learn from different environments. In other words, *we and we alone are the arbiters of our own minds*; they are only what our life experiences and opportunities enable them to be.

In contrast, according to the evolutionary science model, the assembly of the mind is certainly affected by variation in environment/learning experience and profoundly so. But it is also – because of genetic inheritance, through Darwinian selection in the ancestral past – partially and variably structured at birth. From this pre-structure, then, the mind develops a degree of prepared learning. 'In prepared learning, we are innately predisposed to learn and thereby reinforce one option over another. We are "counterprepared" to make alternative choices, or even actively to avoid them' (Wilson 2012). Humans learn then in ways that are modulated in part by innate predispositions and by partially instinctual responses to certain environmental cues that influence a range of variation in particular impulsive motivations and behaviours that would have been important in affecting the reproductive success of ancestors. This conclusion is supported by decades of research on families with twins and adoptees, showing that genetics accounts for about 50% of most human psychological variation (Plomin 2018).

Some features of human behaviour are strongly determined by genes. For example, *Homo sapiens* uses complex language today, and this was also true of our ancestors going back many thousands of years. But it was not true for all of our predecessors or any of them from our very distant past. They did not possess the required genetic mutations. So important was this capacity in effecting evolutionary fitness that it became virtually a human universal and of course is not shared today with any of the other Great Apes. This between-species variation in complex language capacity – and its profound role in shaping human culture – evolved only because first there was within-species genetic variation for natural selection to act upon in early humans. In other words, origination of new species (with diagnostic characteristics) first requires genetic variation (for these same characteristics or intermediate forms) *within* ancestral species. Accordingly, most of our behaviours and cultures differ dramatically from other Great Apes not primarily because of the effects of different social learning environments and experiences but because of the expressions of different resident genes (and their interactions with social learning environments and experiences; see below). And importantly, this resulted because one or a few of our distant predecessors had acquired the original defining mutations for these gene variants (or their precursors), while their contemporary fellow humans had not. *Only the former, not the latter, became our ancestors.* The same effects of genetic variation are at work today within *Homo sapiens*, and the same force of natural selection is operating – affecting what our descendants will be, just as it has affected what we are.

Another example of a human universal is represented by the cross-cultural attraction to the rhythmic and melodic structure of music, and the emotions that it triggers, especially in groups and especially when combined with dance. This has been widely interpreted as having ancestrally adaptive social roles in advertising affective states, promoting social affiliation, communication and attachment behaviours, and in announcing and solidifying group identities and coalitions (Curtis and Bharucha 2010; Ukkola-Vuoti et al. 2011; Fitch and Popescu 2019; Mehr et al. 2019). Dunbar (2004), asks:

> Were dance and singing, and perhaps the rhythmic clapping of hands that so often accompanies both of these an early supplement to physical grooming that allowed Homo erectus to enlarge its groups beyond the limit imposed by the immediate time constraints of grooming? ... something similar to non-human primate contact calling must have bridged the gap between the first rise in group size above the conventional non-human primate limit (about 60-70 individuals) and the rise of true language (once group size had exceeded around 120). Given what we know both about primate contact calls and their use in choruses and about music, it seems increasingly likely to me that it was singing that filled this gap. ... Of particular significance here is the fact that we can induce these emotional effects by music alone without the need for any words. Wordless songs and the pure tonalities of musical instruments produce the same effects as the most rousing lyrics. Gregorian plainsong of the Catholic monastic tradition provides a particularly obvious example of this. It is the sounds of harmonious chant that we find so compelling, not the words — especially given the fact that most of it is in ancient Latin and not understood.

Recent research has shown that variation in musical expression/talent is a product of not just social learning (e.g. from music lessons); it also has a conspicuous genetic component[1] (Stetka 2014). But this can also interact with social learning, e.g. if a child is naturally talented at something (e.g. music or reading), she/he is also more likely to be naturally attracted to – and to enjoy – practicing it, i.e. even without being prodded by parents (Mitchell 2018). Similarly, children whose parents read to them may be better readers at school, but variation in how much parents read to their children might also be a response to intrinsic variation in how much the children enjoy reading. The latter may be a product of inherited genetic variation affecting how much their parents enjoy reading and/ or a product of children (who like to read or be read to) using their environment to feed their intrinsic appetite for reading – by asking their parents to read to them (Plomin 2018). In other words, both genes and exposures to environments can affect our motivations, but genes can also affect which environments we prefer (Briley and Tucker-Drob 2013).

One's favourite musical genre is determined largely by what one has had opportunity to hear, e.g. the style that one grew up with or learned in the years of youth, and this can vary widely in distinguishing cultural group variation, both regionally and across generations (Offord 2017). A broadly popular music culture, however

---

[1] In the next chapter we explore how such talent probably provided a 'fitness signal' – advertising a high-quality mind in our ancestors that was attractive to potential mates.

(e.g. jazz, rock, country, hip hop), is sustained, I suggest, not just by chance, i.e. not just from a collective momentum of many people haphazardly imitating and responding to a particular musical genre. Historically, the most popular local music cultures 'of the day' were probably also informed in part by inheritance of 'whispering genes' that inspired ancestors to identify with these genres because they evoked adaptive emotions (in both the artists who created them *and* their fans who enjoyed them), e.g. associated with feeling a sense of 'membership' in something larger than 'self' (i.e. a local social or peer group) or associated with signalling status or prestige, thus promoting ancestral gene transmission through successful mate acquisition (see Chap. 7) or through domains of leisure (see Chap. 9) or domains of legacy (see Chap. 10). Genes and environments then normally interact in affecting not just individual behaviours but also many of the 'collective behaviours' that we call human culture, i.e. the collection of 'memes' (knowledge, ideas, values, and beliefs) that reside in the minds of individuals belonging to a local population and that manifest as particular behaviours, habits, customs, styles, tools, techniques, procedures, and traditions that are copied and transmitted between (and within) generations as a result of imitation, communication, teaching, and learning, and that differ noticeably in character from the memes, teachings, and traditions of other social groups.

Genetic Versus Memetic Legacies

> It is my belief that the situation is ripe for a synthesis. There is a feedback relationship between the biological and the cultural evolutions of mankind. The big problem is how this relationship operates and where it is taking the human species. Let no one mistake it — there are no easy answers here. The matter needs careful rethinking in the light of the present knowledge, and even more, it needs further research, research in which biologists who are not oblivious of their limitations should cooperate with social scientists who are not blind to theirs. (Dobzhansky 1963)

Human behavioural traits that are common today include those of our predecessors that left the most genetic descendants or 'genetic legacy' – because natural selection rewarded the transmission success of some of their particular resident genes that informed adaptive thinking, decisions, and actions. This of course is biological evolution, where the mechanism for genetically transmitted inheritance involves sexual reproduction. Some of these traits are essentially human universals – 'hard-wired' (genetically determined) behaviours – because they were core determinants of reproductive success in our ancestral past (e.g. hunger, fear, sexual arousal, potential for spoken language).

Other features of contemporary human behaviour include those of our predecessors that left the most 'memetic legacy' – because cultural selection rewarded socially transmitted inheritance of particular resident memes that characterized their behaviours. This then is cultural evolution, where the transmission mechanism involves social learning (oral and written) – without necessarily involving any change/variation in gene frequencies. Examples of apparent 'cultural determinism' may include variation in the 'national' (favoured) sport of different countries (e.g. football, baseball, ice hockey) and variation in the syntax, grammar, dialects, vocabularies, accents, jargon, slang, and colloquialisms of different languages (e.g. English versus Mandarin versus Spanish, today).

Jukeboxes and Colouring Books

One's experience of environmental variation can also modulate gene expression by affecting which genes are turned 'on' or 'off'. In other words, gene expression in many cases takes its cue from the environment, including the sociocultural/economic environment – routinely in ways that rewarded the reproductive success of ancestors. The mind then is not like a 'blank slate', with features solely determined by what gets learned or 'added to it' from outside inputs. Nor is the mind like a blueprint or a computer algorithm, with features determined solely by 'coding' from gene expression. There is very little, therefore, about the roles of genes and environment in the mental life of humans that reflect alternatives in a 'tug of war' (famously referred to as 'nature versus nurture'). In other words, their relationship in human evolution has been more of an inter-dependence or 'blending' and not where one has generally overwhelmed the other (Mitchell 2018; Plomin 2018). Dobzhansky (1963) understood this, over half a century ago:

> The premise which cannot be stressed too often is that what the heredity determines are not fixed characters or traits but developmental processes. The path which any developmental process takes is, in principle, modifiable both by genetic and by environmental variables. It is a lack of understanding of this basic fact that is, it can safely be said, responsible for the unwillingness, often amounting to an aversion, of many social scientists … to admit the importance of genetic variables in human affairs.

The human mind then resembles a jukebox (Tooby and Cosmides 1992), where a number of 'tracks' (genetic instructions) are already stored in the 'machine', but particular biophysical/chemical and social environments determine which 'buttons' get pressed to play particular 'tracks'. The human mind can also be likened to a colouring book (Kenrick et al. 2010), where the inner structure of pre-drawn lines (genetic instructions) interacts with environmental inputs (different artists with differently coloured crayons) to determine the final phenotype of the behaviour (picture) (Box 6.1). From Dobzhansky and Montagu (1947): 'The process of natural selection in all climes and at all times have favored genotypes which permit greater and greater educability and plasticity of mental traits under the influence of the uniquely social environments to which man has been continuously exposed'.

Humans then are susceptible to social or cultural influence, as Nettle (2009) put it, '… because they have evolved mechanisms that make them so'. Tomasello (2014) distinguishes several layers of socio-cognitive skills – involving shared, joint, or collective intentionality – each of which requires ontogenetic development that is contingent upon the availability and effectiveness of learning within enriched sociocultural environments. In other words, these skills ' … are not simply innate, or maturational; they are biological adaptations that come into existence as they are used during ontogeny to collaborate and communicate with others' (Tomasello 2014).

Box 6.1: The 'Colouring Book' Metaphor of the Human Mind

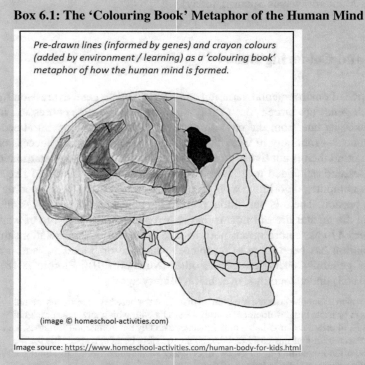

Pre-drawn lines (informed by genes) and crayon colours (added by environment / learning) as a 'colouring book' metaphor of how the human mind is formed.

(image © homeschool-activities.com)

Image source: https://www.homeschool-activities.com/human-body-for-kids.html

Image source: https://www.homeschool-activities.com/human-body-for-kids.html

Adaptive behaviours then are commonly said to be 'evoked' by gene expression that is contingent upon environment or is 'context sensitive', e.g. depending on age, physiological stage, or developmental induction (from early life inputs or conditions). Context sensitivity requires that, in the ancestral past, '… each of the different environmental states has been experienced recurrently, and the best behavioural strategy to follow for a given environmental state has recurrently been the same' (Nettle 2009). In many cases, therefore, collective behaviour – culture – is 'evoked', i.e. where genes are involved not in the determination of culture but in the 'meta-determination' of culture. In other words, gene expression provides the cognitive scaffolding on which particular features of human motivations and cultures are shaped and built through a variety of potential experiences with ecological and social environments. Nettle (2009) elaborates:

> In some situations, one [cultural] phenotype is advantageous, and in other situations, another. What selection builds in this instance is a mechanism for seeking and internalizing cues of which of the possible environmental states obtains locally, and calibrating the [cultural] phenotype accordingly. ... The information on how to build each possible [cultural] phenotype is built into the organism by selection, as is the menu of which cue should evoke which phenotypic state. The role of the current environment here is to provide the cues ... Not only was the human crucible of Pleistocene Africa very temporally labile, with repeated rapid alternations between humidity and aridity, but humans were using several different niches within it, and for tens of thousands of years humans were constantly changing habitats as they colonized new areas (Wells and Stock 2007). Thus, there is abundant scope for humans to have developed the kind of environmentally contingent behaviors that are subserved by evoked-culture adaptations.

Culture as an Adaptation

Human culture then is commonly adaptive simply because traditions and tools commonly enhanced ancestral genetic fitness. And an important advantage of cultural evolution is that it can occur much more quickly than biological evolution (Box 6.2). From Richerson and Boyd (2005):

> Culture is adaptive because populations can quickly evolve adaptations to environments for which individuals have no special-purpose, domain-specific, evolved psychological machinery to guide them. ... In the wildly varying environments of the Pleistocene, individuals were better off relying upon fast and frugal learning heuristics to acquire pretty good behaviours 'right now' rather than await the perfect innate or cultural adaptation to an environment that would be gone before perfection could evolve.

And from Nettle (2009):

> Social learning – copying the behaviour of others in the population – can be selected for when the costs of individual learning are substantial and there is some cross-individual and cross-time consistency in which behaviours are optimal. However, social learning is under frequency - dependent selection and thus at equilibrium there will always be some mix of individual experimentation and social learning. Whenever there is some bias in who is copied (e.g. differential copying of those who are most successful) then social learning will lead to locally adaptive behaviour much of the time.

Box 6.2: Cultural Evolution Can Occur More Quickly Than Biological

When environmental change is slower, genetic evolution can track it more closely; i.e. with adaptive traits, informed by effects of genes, favoured in their copying and transmission success to future generations .

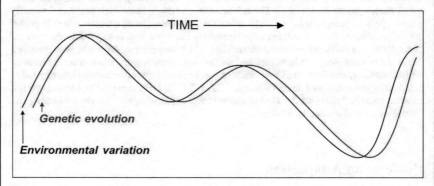

When environmental change is faster, genetic evolution often cannot 'keep up', but cultural evolution commonly can. As Richerson et al (2010 put it: *"As cultural adaptations became important, much could be gained from imitating a seemingly successful idea or practice. If people can judge what is successful, or who is successful, new adaptive variation can rapidly spread through an entire population, sometimes within one generation."*

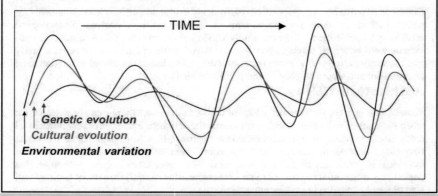

Natural selection therefore can be expected to favour both 'smart innovators' and 'smart copiers' (Laland 2017), each depending on balances between costs and benefits (Box 6.3). With an increase in 'trial and error' effort, innovators should be expected to benefit from greater acquisition of fitness, promoting, new discoveries, but balanced by an increasing potential cost of lost time/effort for engaging in (and gaining benefits from) some social learning. Similarly, with greater 'information scrounging' effort, copiers should be expected to benefit from acquiring cheap (free) fitness promoting discoveries made by innovators, but this is also balanced by an increasing potential cost of accumulating outdated, inaccurate, or irrelevant information, plus an increasing potential cost of lost time/effort for engaging in (and gaining benefits from) some innovation 'on the side' (Box 6.3).

'Smart innovators' then have flexible deployment of innovative tactics, by using cues from the environment (affecting the cost and benefit functions) to track opportunities for new discovery effectively and thus switching strategically between fruitful 'knowledge producing' activities and selective 'knowledge scrounging' activities (to take advantage of the efforts of other innovators). Similarly, 'smart social learners' take cues from the environment which enable recognition of outdated, inaccurate, or irrelevant information and hence adjustment of copying effort accordingly (Box 6.3). Importantly, in the modern 'online' digital age, with multiple and rapidly changing domains for 'search engines' and social media, the potential pace of these trackings and adjustments are ramping up exponentially – with potential consequences for cultural evolution (and possibly biological evolution) that we are only beginning to understand (Acerby 2020).

Box 6.3: Smart Innovators and Smart Cultural Learners

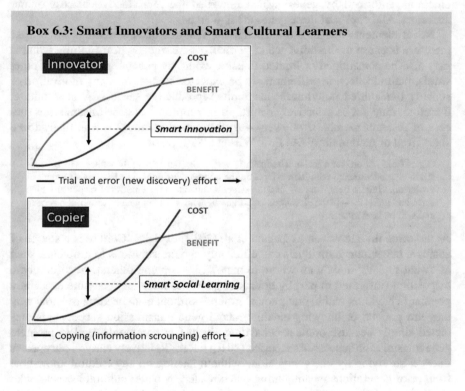

Why Have Only Certain Cultures Evolved and Not Other Imaginable Ones?

Consider, for example, the famous seven 'deadly (cardinal) sins' from Dante's *Inferno*, written about 700 years ago: sloth (apathy), envy, gluttony, lust, pride (vanity), revenge (wrath), and greed. What are the roots of these so-called human 'weaknesses'? – cultural transmission from social learning? – supernatural forces, e.g. satanic temptations, the conflict between 'good' and 'evil'? The evolutionary science model would predict that these universal themes within our cultural codes of conduct exist in their particular form because they are reflections of innate human motivations (Saad 2007). In other words, the reason why these 'deadly sins' are so common, alluring, and/or difficult to eradicate – and so have inspired cultures of art, music, film, and literature – is probably because they characterize motivations and emotions, informed by genes, which rewarded the reproductive success of our ancestors. And copies of these genes are now in us.

Novel elements that define characteristic features of new cultures often arise therefore because of the influence of particular genes/genotypes. Cultural features may also be constrained or limited to some extent by genetic influences. In other words, natural selection will normally be expected to disfavour any motivation or socially transmitted inheritance that limits reproductive success. A new culture, therefore, may not be selectively neutral; if it compromises gene transmission success of resident members/followers – a 'Darwinian Paradox' – it is likely to be short-lived or marginalized. As E.O. Wilson (1978) put it:

> Genes hold culture on a leash. The leash is very long, but inevitably values will be constrained in accordance with their effects on the human gene pool. The brain is a product of evolution. Human behaviour — like the deepest capacities for emotional response which drive and guide it — is the circuitous technique by which human genetic material has been and will be kept intact.

At the same time however, as Laland et al. (2010) explain: 'Culture is a source of adaptive behaviour; individuals can efficiently acquire solutions to problems, such as "what to eat?" and "with whom to mate?", by copying others'. Accordingly, a disposition (informed in part by effects of genes) to learn and copy the especially popular behaviours within one's social group – without even needing to know why they are popular or knowing anything about gene transmission success – is predicted simply because popular behaviours do in fact commonly tend to promote gene transmission success. As Henrich (2016) put it: 'Relatively early in our species lineage, surviving by one's wits alone without leaning on any cultural know-how from prior generations meant getting outcompeted by better cultural learners, who put their efforts into focusing on what and from whom to learn'.

An outstanding innovator might be a stupid social learner or a brilliant one, but natural selection will favour the latter, which probably accounts in large measure for our species' success. It was theory of mind that enabled us to become superb 'copy-cats', and hence, together with evolution of complex language, social learning ramped up the pace of cultural evolution and its blending with genetic evolution in shaping our evolved psychologies (Wood 2020). And when the culture of written language evolved from spoken language, the culture of reading – especially literary fiction it seems, according to recent research – may have in turn bolstered mind-reading skills (Kidd and Castano 2013) and thus a progressive sophistication of smart innovation blended with smart copying.

Biocultural Evolution

Gene pools (biology) and meme pools (culture) therefore tend to coevolve in a kind of self-sustaining (or self-adjusting) feedback relationship (Box 6.4), with genes holding culture (socially transmitted inheritance) 'on a leash' and, at the same time, culture holding genes (genetically transmitted inheritance) 'on a leash'. In other words, any genes/genotypes that limit smart social learning (within the prevailing local background ecological and social environment) are likely to be rare (culture holding genes on a leash), and any learned culture/memes that limit reproductive (gene transmission) success (within these prevailing local environments) are also likely to be rare (genes holding culture on a leash).

Cartwright (2008) explains with an example: 'In early traditional cultures, there was probably a large degree of synergy between genes and memes and this may persist into modern times. The linked memes in Catholicism that restrict birth control and also insist that offspring are raised in the faith, for example, have the dual effect of increasing the spread of memes and the genes of those professing the memes'. Some of those genes are likely to inform adaptive (i.e. 'gene-transmission generating') behaviours involving inclinations to follow instructions through blind faith in particular local traditions (like religions) learned during youth – and involving attraction to religiosity in general (a topic examined in more detail in Chap. 10).

In the 'landscape' of biocultural evolution, therefore, 'evolved psychology' and 'evolved culture' are normally blended (Box 6.4). Specifically, an adjusted meme pool (resulting from new social learning or use of new technology informed by memes) – together with genetic novelty generated by mutation, gene flow, and genetic drift – modulates the effects of natural selection in favouring reproductive (gene transmission) success (within the prevailing local ecological and social environments) for certain genotypes that influence neurobiology (brain structure and biochemistry), in ways which evoke certain phenotypes of cognition, intelligence, imagination, language, desires, emotions, drives, and decisions, thus triggering change in local phenotypes of behaviours and actions. This biological evolution (adjusted gene pool) in turn modulates/informs the effects of local cultural selection

Box 6.4: The Landscape of Biocultural Evolution

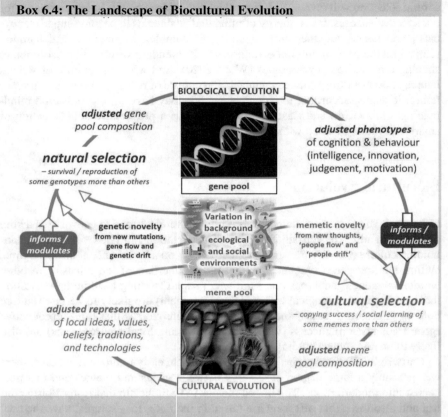

Image sources: https://www.shutterstock.com/image-vector/cartoon-map-seamless-pattern-river-187438748;https://alessandria.today/2021/11/14/la-forza-della-parola-la-parola-come-strumento-di-giovanna-fileccia/

in favouring transmission (social learning) success of certain memes over others, again within the prevailing local ecological and social environments. Variation in the latter also introduces 'memetic novelty' (analogous to genetic novelty), resulting from haphazard generation of new thoughts and ideas (analogous to new gene mutations) within resident minds or from 'people flow' or 'people drift' (analogous to gene flow and genetic drift respectively), bringing new memes subjected to cultural selection and hence with potential contribution to local meme pool adjustment. These cumulative effects on cultural evolution, again, ratchet up its modulating effects on natural selection, favouring (again) certain genes/genotypes (e.g. associated with smarter brains, more complex language), thus sustaining a prevailing culture or generating (again) a different/more sophisticated culture (Box 6.4). Importantly in this 'landscape', background environments are shaped not just by fortuitous physical, abiotic forces but also by intentional behaviours and actions of resident individuals themselves.

The cultures of fictional narrative, e.g. literature, film, theatre, and musical lyrics, represent obvious examples of the potential for cultural selection through social learning. As Carroll (2018) explains: 'They expand and refine readers' imaginative understanding of life, thus helping them organize their values and beliefs. The imaginative virtual worlds of fiction penetrate deeply into personal identity, inspiring visions that have sometimes had world-historical consequences ... Fictional stories provide readers with simulated experiences entirely outside the range available to any one person dependent on his or her own direct contact with people and places ... Reading fiction puts one in communication with strong personalities, minds that have highly developed powers of observation and reflection, and imaginations capable of evoking powerful passions and subtle sensations ... Responding to those minds, both sympathetically and critically, is a form of education, and education is itself a motive rooted in the cultural capacities that are part of the adaptive repertory of *Homo sapiens*'.

Clever minds and clever cultures thus coevolved. In other words, while human minds are 'inventing' cultures, cultures are also 'inventing' human minds (Stewart-Williams 2018). Accordingly, new cultures arise not just because of random firing of neurons generating new ideas/traditions (memes) that are copied and transmitted through social learning. They arise also because of local genetic change (biological evolution) affecting cognitive traits that inform intrinsic skills, motivations, and intentional actions and behaviours (including 'smart innovation' and 'smart copying') that in turn influence the process of cultural selection that shapes the composition of local meme pools (Box 6.4).

Over time, therefore, the products of cumulative cultural evolution influence the composition of local gene pools by favouring genotypes that inform minds equipped with greater intelligence, creativity, computational power, and capacity for using complex language and hence more culture evolution. The evolution of complex language skill in particular must have ramped up the historical pace of biocultural evolution by orders of magnitude. From Morrison (2020): 'Under building pressure for an enhanced signaling system capable of supporting joint attentional-intentional activities, a cultural tradition of disambiguated indexical pointing (a finger point disambiguated by a facial expression, vocalization, or other gesture), combined with increasingly sophisticated mindreading circuitry and prosocial tendencies, may have sparked the first in the series of biocultural explosions that led from a simple protolanguage to fully modern human language'.

Through biocultural evolution, therefore, our species became more and more capable of ever more (cumulatively) sophisticated cultures. In each generation, those of our ancestors who were more adept at acquiring, responding to, and developing culture (e.g. stone tools, control of fire, cooking, agriculture, writing, religion, etc.) generally had a distinct advantage over those who were less adept, i.e. they transmitted copies of their genes at a greater rate into future generations. 'We are cultural inheritors of Darwin, Beethoven, Newton, Plato, the ancient nameless inventors of agriculture and of fire, most of whom were not our biological ancestors.

Mankind's cultures, like its genes, have evolved and are evolving' (Dobzhansky 1964). In other words, as Vince (2020) put it: 'We are making ourselves'. From Dobzhansky (1961): 'Man adapts his environments to his genes more often than he adapts his genes to his environments. But, as pointed out, the two methods of adaptation are complementary and not alternative'. Hence, at the same time, 'For good or for ill, natural selection fits man to live in the environments created by his own culture and technology' (Dobzhansky and Allen 1956). And from Flinn and Alexander (1982):

> We think the evidence suggests that cultural traits are, in general, vehicles of genic survival, and that the heritability of cultural traits depends on the judgments (conscious and unconscious) of individuals with regard to their effects on the individual's inclusive fitness.

Most (if not all) social and cultural norms then probably have a foundation, at least partially, in the effects of Darwinian natural selection on human thinking, motivations, emotions, and behaviours. And these norms can in turn act as agents of natural selection, driving feedback effects on the transmission of genes that further influence human dispositions and cultures (Richerson et al. 2010). Dobzhansky (1963) described it over half a century ago:

> … since the environment in which man lives is in the first place his sociocultural environment, the genetic changes induced by culture must affect man's fitness for culture and hence may affect culture. The process thus becomes self-sustaining. Biological changes increase the fitness for, and the dependence of their carriers on culture, and therefore stimulate cultural developments; cultural developments in turn instigate further genetic changes. This amounts to a positive feedback relationship between the cultural and the biological evolution.

Humans, for example, created the culture of how to live in dense crowds. And a reference to Charles Galton Darwin (1952) from Chap. 5 serves here again as an illustration of probable feedback effects on biological evolution: 'Life in the crowded conditions of cities has many unattractive features, but in the long run these may be overcome, not so much by altering them, but simply by changing the human race into liking them'.

Can We Will What We Want?

Genes can '… hold culture on a leash' (Wilson 1978). But, as examples above illustrate, culture can also 'hold genes on a leash'. Barash (2000) captures the balance: 'There is very little in the human behavioral repertoire that is under genetic control, and very little that is not under genetic influence'. Put another way, according to German philosopher Arthur Schopenhauer (1788–1869): 'A human can very well do what he wants, but cannot will what he wants' (quoted from Overbye 2007; Smith 2013). As Plomin (2018) describes it: '… we can turn the television on or off as we please, but turning it off or leaving it on pleases individuals differently, in part due to genetic factors. Genetics is not a puppeteer pulling our strings. Genetic influences are probabilistic propensities, not pre-determined programming'.

'Free will' then is probably (or at least usually) an illusion, i.e. much of the thinking that guides our actions is outside of our consciousness (Critchlow 2019). Accordingly, in many instances, we do not 'will' how our brain thinks or what actions result from its instructions. Consider, for example, how, when operating a vehicle, driving actions often routinely result from impulsive, adaptive, safety-preserving motivations that one is not consciously thinking about – with no recollection of the actions, or the 'decisions', or even the landscape that was travelled in the process.

The common inclination of our species to assume that we routinely have 'free will', it seems to me, is just an extension of our deeply ingrained delusion that we have an 'inner self' – a mental life that gives us complete agency over all that we think, all that we do, and all that we are (Chap. 4) – and that we can somehow leave a symbolic legacy of this 'self' for posterity, even while knowing with certainty that our material existence will someday end. Importantly, our psychology for these sentiments coevolved with our cultures that reinforce them (see Chap. 10). If we can be so easily fooled and manipulated (by our psychologies and cultures) into believing in this, then it should be no surprise that we routinely convince ourselves that we can 'will what we want'. The reality, however, is, mostly, we can't.

Importantly though (as this book and many others illustrate), our uniquely human cognitive capacities allow us to study and understand the potential influence of genes on our behaviours and the consequences of these behaviours and so enable us (if we intentionally choose) to exercise control over whether we will follow certain instincts and inclinations that are informed by effects of genes. In other words, in this respect, we *can* in fact do what we want – in the sense that we can choose to say 'no' to our genes and follow instead a culture that advises us to do so. Our genes 'do not bark orders'; they only 'whisper to us' (Barash 2014). Listening and following some of these 'whispers' – filtered and modulated by human decisions that allow but nevertheless constrain their influence (i.e. with culture 'holding genes on a leash') – led to motivations that rewarded the reproductive success of ancestors and thus played a role in shaping/evoking some of the features of cultures (e.g. legal systems that prohibit the mistreatment of fellow citizens). Additionally, certain genes/alleles evoking different 'whispers' – residing in other predecessors that followed these 'whispers' – were less successful in rewarding reproductive success. These latter genes/alleles then never (or only rarely in low frequency) got transmitted to descendants. Hence, some potential alternative cultures never evolved, or they manifest only rarely/temporarily (e.g. 'suicide cults'), i.e. with genes 'holding culture on a 'leash'.

A Postlude on Nature Versus Nurture

There is a long history of controversy in the evolutionary interpretation of human behaviour and culture going back to Darwin himself (Darwin 1871, 1872). Even in *The Origin*, Darwin (1859) writes: 'In the distant future I see open fields for more important researches. Psychology will be based on a new foundation, that of the necessary acquirement of each mental power and capacity by gradation'. Darwin knew that this was a delicate topic and potentially dangerous, as was borne out in the Social Darwinism and eugenics movements that developed in subsequent decades. For some time, therefore, most researchers interested in the biological basis of behaviour preferred to steer clear of human study material. A growing knowledge of genetics early in the last century, however, brought new life to Darwinism – the 'modern evolutionary synthesis'. And one of its main architects, Theodosius Dobzhansky, was drawn, undeterred, to the application of Darwinism in search of a deeper understanding of the human condition. Citations representing this pioneering influence of Dobzhansky thus figure prominently throughout the present volume (Dobzhansky and Montagu 1947; Dobzhansky and Allen 1956; Dobzhansky 1961, 1962, 1963, 1964, 1967, 1973).

But it was not until the 1970s that this application of Darwinism (despite continuing controversy) started to become a conspicuous focus of study, emerging from several perspectives, variously called 'human sociobiology', 'human behavioural ecology', 'evolutionary psychology', 'evolutionary sociology', 'cultural evolution theory', and 'gene-culture coevolution' (see Brown et al. (2011) for a review). These differ in emphasis on certain details concerning levels of importance for and degrees of interaction between genetically and socially transmitted inheritance and concerning the roles of environmental factors in modulating behavioural development and in eliciting alternative behaviour patterns. Debates regarding these details can be interesting for academics and fruitful for theory maturation, but a lot of it is about 'turf protection' – bickering generated from theory tenacity and confirmation bias (see Chap. 12). For both the student and the general public, it is more important to have a firm grasp of the common ground in these various points of view, represented in the central underlying theme of this book: *Humans have motivations (needs and drives) – manifesting in their emotions, behaviours, and cultures – shaped by effects of natural selection on gene pools (biological evolution) and by the effects of cultural selection on meme pools (cultural evolution) but also by the modulating effects of cultural evolution (memetic inheritance) on natural selection and the modulating effects of biological evolution (genetic inheritance) on cultural selection* (Box 6.4).

Details of these motivations, needs, and drives – and their relationships with cultural evolution – are the subjects of the next five chapters, in preparation for the last two chapters. There, we explore how – as Flinn (2013) predicts (I believe, accurately) – 'Understanding the evolutionary basis for human culture is far more than a difficult academic issue. Such an understanding is critical to solving humanity's great problems of environmental degradation, social injustice, over-population, and war'.

The debate over 'nature' versus 'nurture' has languished long, as Pinker (2004) recounts:

During much of the twentieth century, a common position in this debate was to deny that human nature existed at all — to aver, with José Ortega y Gasset, that 'Man has no nature; what he has is history.' The doctrine that the mind is a blank slate was not only a cornerstone of behaviorism in psychology and social constructionism in the social sciences, but also extended widely into mainstream intellectual life.

It is interesting to speculate, however, that the 'blank slate' model of the mind, *itself*, may be a cultural (memetic) product of evolution by natural selection. Being a staunch defender of the 'blank slate' may just be a modern manifestation of a deeply ingrained disposition (inherited from the ancestral past) to be a staunch defender of the *'self'* – a disposition that was adaptive to ancestors because it helped to dispel the worry that one might be stuck with an intrinsically inferior 'self', prone to impermanence. The evolutionary science model of the mind poses a threat, therefore, because it espouses that an 'inferior self' is likely to be informed, at least partially, by less-than-superior genes. Tenacious belief in the 'blank slate' then (like belief in religion) provides a buffer from the fear of having limited potential for memetic legacy (with the latter conjured by self-impermanence anxiety; see Chap. 4 and explored further in Chap. 10). Like belief in free will (discussed above), belief in the 'blank slate' also facilitates the appealing notion of a sense of 'ownership' of who you are, i.e. giving license for *you* – your 'inner self' rather than your impersonal, material DNA – to take credit for the kind of person you turned out to be and hence the quality and perpetuity of memetic legacy that you have potential to leave, thus bolstering self-esteem. If this interpretation is correct, then the debate between the standard social science ('blank slate') model of the human mind and the evolutionary science model of the human mind is misguided; in other words, the 'blank slate' view is not in conflict with Darwinian evolution – it is a cultural product of it. As Hopcroft (2016) put it: '… without the human biological predisposition toward sociality, there would be no need for a discipline [social science] that focuses on the emergent properties of the human social group'.

What does it mean to say that human behavior has no genetic cause because it depends on society. … Society is not an alternative to genetic programming. It requires it. To become a member of any kind of society, an infant must be programmed to respond to it. Others give him cues. But he has to be able to pick them up and complete the dialogue. … Babies do not only need to suck and cry to survive. A baby that fails as time goes on, to smile and talk, laugh and weep, to meet the eyes of those around it, to seek and follow its parent, to treat those around it with affection, to want their company and approval, to play and explore the world, cannot join its society. But these proceedings are thoroughly typical of those by which helpless young in other species secure the care and attention of their elders and are integrated into their society. They are normal patterns in social animals. In the evolutionary perspective, then, society and genetic programming imply each other. And the more complex the society, the richer the genetic programming has to be. (Midgley 1979)

References

Acerby A (2020) Cultural evolution in the digital age. Oxford University Press, Oxford
Barash DP (2000) Evolutionary existentialism, sociobiology, and the meaning of life. BioScience 50:1012–1017
Barash DP (2014) Buddhist biology: ancient eastern wisdom meets modern western science. Oxford University Press, Oxford
Briley DA, Tucker-Drob EM (2013) Explaining the increasing heritability of cognitive ability across development: a meta-analysis of longitudinal twin and adoption studies. Psychol Sci 24:1704–1713
Brown GR, Dickins TE, Sear R, Laland KN (2011) Evolutionary accounts of human behavioural diversity. Phil Trans R Soc B 366:313–324
Carroll J (2018) Minds and meaning in fictional narratives: an evolutionary perspective. Rev Gen Psychol 22:135–146
Cartwright J (2008) Evolution and human behaviour: Darwinian perspectives on human nature, 2nd edn. MIT Press, New York
Critchlow H (2019) The science of fate: why your future is more predictable than you think. Hodder & Stoughton, London
Curtis ME, Bharucha JJ (2010) The minor third communicates sadness in speech, mirroring its use in music. Emotion 10:335–348
Darwin C (1859) On the origin of species. Facsimile of the first edition. Harvard University Press, Cambridge
Darwin CR (1871) The descent of man, and selection in relation to sex. John Murray, London
Darwin CR (1872) The expression of the emotions in man and animals. John Murray, London
Dobzhansky T (1961) Man and natural selection. Am Sci 49:285–299
Dobzhansky T (1962) Mankind evolving: the evolution of the human species. Yale University Press, New Haven
Dobzhansky T (1963) Anthropology and the natural sciences: the problem of human evolution. Curr Anthropol 4:138+146–148
Dobzhansky T (1964) Heredity and the nature of man. New American Library, New York
Dobzhansky T (1967) The biology of ultimate concern. The New American Library, New York
Dobzhansky T (1973) Ethics and values in biological and cultural evolution. Zygon 8:261–281
Dobzhansky T, Allen G (1956) Does natural selection continue to operate in modern man? Am Anthropol 58:592–604
Dobzhansky T, Montagu MFA (1947) Natural selection and the mental capacities of mankind. Science 107:587–590
Dunbar R (2004) The human story: a new history of mankind's evolution. Faber and Faber, London
Fitch WT, Popescu T (2019) The world in a song. Science 366:944–945
Flinn M (2013) Biology and culture. In: Summers K, Crespi B (eds) Human social evolution: the foundational works of Richard D. Alexander. Oxford University Press, New York
Flinn MV, Alexander RD (1982) Culture theory: the developing synthesis from biology. Hum Ecol 10:383–400
Henrich J (2016) The secret of our success: how culture is driving human evolution, domesticating our species, and making us smarter. Princeton University Press, Princeton
Hopcroft RL (2016) Grand challenges in evolutionary sociology and biosociology. Front Sociol 1:1–3
Kenrick DT, Nieuweboer S, Buunk AB (2010) Universal mechanisms and cultural diversity: replacing the blank slate with a coloring book. In: Schaller M, Norenzayan A, Heine SJ, Yamagishi T, Kameda T (eds) Evolution, culture and the human mind. Psychology Press, New York
Kidd DC, Castano E (2013) Reading literary fiction improves theory of mind. Science 342:377–380
Laland KN (2017) Darwin's unfinished symphony: how culture made the human mind. Princeton University Press, Princeton

Laland KN et al (2010) How culture shaped the human genome: bringing genetics and the human sciences together. Nat Rev Genet 11:137–148

Mehr SA, Singh M, Knox D et al (2019) Universality and diversity in human song. Science 366(6468):eaax0868. https://doi.org/10.1126/science.aax0868

Midgley M (1979) Beast and man: the roots of human nature. The Harvester Press, Hemel Hempstead

Mitchell KJ (2018) Innate: how the wiring of our brains shapes who we are. Princeton University Press, Princeton

Morrison DM (2020) Disambiguated indexical pointing as a tipping point for the explosive emergence of language among human ancestors. Biol Theory 15:196–211

Nettle D (2009) Beyond nature versus culture: cultural variation as an evolved characteristic. J R Anthropol Inst 15:223–240

Offord C (2017) Understanding the roots of human musicality. The Scientist, 1 Mar 2017. https://www.the-scientist.com/features/understanding-the-roots-of-human-musicality-31917

Overbye D (2007) Free will: now you have it, now you don't. New York Times, 2 Jan 2007. http://www.lifesci.utexas.edu/courses/THOC/Readings/Overbye_Free%20Will%20NYT2007.pdf

Pinker S (2004) Why nature and nurture won't go away. Daedalus 133:5–17

Plomin R (2018) Blueprint: how DNA makes us who we are. MIT Press, Cambridge

Richerson PJ, Boyd R (2005) Not by genes alone: how cultural transformed human evolution. University of Chicago Press, Chicago

Richerson PJ, Boyd R, Henrich J (2010) Gene-culture coevolution in the age of genomics. Proc Natl Acad Sci 107:8985–8992

Saad G (2007) The evolutionary bases of consumption. Lawrence Elbaum Associates, London

Smith E (2013) The Thomas Mann handbook - everything you need to know about Thomas Mann. Emereo Publishing, Brisbane

Stetka B (2014) What do great musicians have in common? DNA. http://www.scientificamerican.com/article/what-do-great-musicians-have-in-common-dna/?&WT.mc_id=SA_MB_20140806

Stewart-Williams S (2018) The ape that understood the universe: how the mind and culture evolve. Cambridge University Press, Cambridge

Takacs K (2018) Discounting of evolutionary explanations in sociology textbooks and curricula. Frontiers on Sociology 3:1–14

Tomasello M (2014) A natural history of human thinking. Harvard University Press, Cambridge

Tooby J, Cosmides L (1992) The psychological foundations of culture. In: Barkow J, Cosmides L, Tooby J (eds) The adapted mind: evolutionary psychology and the generation of culture. Oxford University Press, New York, pp 19–136

Ukkola-Vuoti L, Oikkonen J, Onkamo P, Karma K, Raijas P, Järvelä I (2011) Association of the arginine vasopressin receptor 1A (AVPR1A) haplotypes with listening to music. J Hum Genet 56:324–329

Vince G (2020) Transcendence: how humans evolved through fire, language, beauty, and time. Penguin Random House, London

Wells JCK, Stock JT (2007) The biology of the colonizing ape. Yearb Phys Anthropol 50:191–222

Wilson EO (1978) On human nature. Harvard University Press, Cambridge

Wilson EO (2012) The social conquest of Earth. Liveright Publishing Corporation, New York

Wood C (2020) Being copycats might be key to being human. The Conversation, 14 Jan 2020. https://theconversation.com/being-copycats-might-be-key-to-being-human-121932

Chapter 7
The Mating Machine

Every one of our ancestors managed not just to live for a while, but to convince at least one sexual partner to have enough sex to produce offspring. Those proto-humans that did not attract sexual interest did not become our ancestors, no matter how good they were at surviving. (Miller 2000)

Federico Andreotti (1847–1930). (*The Serenade*/Wikimedia Commons/Public Domain)

© The Author(s), under exclusive license to Springer Nature Switzerland AG 2022
L. Aarssen, *What We Are: The Evolutionary Roots of Our Future*,
https://doi.org/10.1007/978-3-031-05879-0_7

Gene transmission and hence fitness, for any animal, is a direct product of sex. Offspring production is the essential currency rewarded by natural selection, and this 'reward' is the principal mechanism that drives biological evolution. It is not surprising therefore that many human behaviours are associated with promoting sexual prolificacy and many popular modern cultural products draw attention to or symbolize features of success in human mating and reproduction. Examples include the music lyrics of the most successful recording artists (Hobbs and Gallup 2011) and the titles and plots of romance novels (Saad 2007; Cox and Fisher 2009). These (as with other cultures explored below and in later chapters) are conspicuous today because they are rooted in our biocultural evolution (Chap. 6)—i.e. our evolved psychologies, shaped by intrinsic motivations and needs that rewarded gene copying and transmission success in our ancestral past, modulated by our evolved cultures, shaped by social learning that rewarded copying and transmission success of complementing memes (Box 6.4).

Determinants of variation in sexual prolificacy can be understood in terms of traditional life history strategy theory, where fitness is affected by a series of trade-offs in the allocation of time, energy, or resources between competing demands/ options (Box 7.1). For example, energy allocated to somatic effort (growth and survival) is unavailable for reproductive effort and vice versa. Similar trade-offs occur between mating effort versus parenting effort, between courtship effort versus mate guarding effort, and between what might be called 'playing-the-field' effort versus 'falling-in-love' effort (Box 7.1).

Natural selection then favours optimal combinations of 'tactics'—represented by different behavioural phenotypes defined by the above trade-offs—that maximize the number of descendants that are left. And the optimal combinations will vary depending on age, socio-cultural/economic environment, and especially biological sex. For example, a man can sleep with 100 different women in a year and father 100 children, but a woman who sleeps with a 100 different men in a year will give birth to only one child (or rarely twins, or triplets). Consequently, prolificacy/fitness in males is generally promoted more by reproductive effort motivations that emphasize quantity of matings (i.e. allocation to mating effort, courtship effort, and 'playing-the-field' effort), whereas prolificacy/fitness in females is generally promoted more by reproductive effort motivations that emphasize quality of mating and offspring care (i.e. allocation to parenting effort, mate guarding effort, and 'falling-in-love' effort) (Box 7.1). Accordingly, young women today tend, on average, to be more interested in 'quality' of sexual experiences, whereas young men are more interested in 'quantity' (Anders and Olmstead 2019).

Box 7.1: Traditional Life-History Strategy Theory

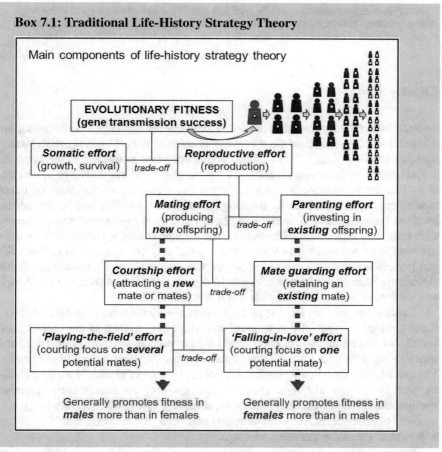

Main components of life-history strategy theory

EVOLUTIONARY FITNESS
(gene transmission success)

Somatic effort
(growth, survival)   trade-off

Reproductive effort
(reproduction)

Mating effort
(producing
new offspring)   trade-off

Parenting effort
(investing in
existing offspring)

Courtship effort
(attracting a new
mate or mates)   trade-off

Mate guarding effort
(retaining an
existing mate)

'Playing-the-field' effort
(courting focus on several
potential mates)   trade-off

'Falling-in-love' effort
(courting focus on one
potential mate)

Generally promotes fitness in
males more than in females

Generally promotes fitness in
females more than in males

Women, therefore, tend to perceive men as less committed in relationships than they really are because the cost of over-perceiving a male's commitment is greater than the cost of under-perceiving commitment. In other words, 'It is much worse to be impregnated by a commitment-pretending psychopath than to doubt a truly committed partner's intentions. Thus, women doubt male commitment, and men feign more commitment than they feel – a never-ending arms race of romantic skepticism and excess that has shaped both female and male mating intelligence' (Geher and Miller 2008). The human mind thus evolved, to a large extent, as a 'mating machine'. Buss (2008) elaborates:

Such tactics require formidable cognitive skills – including mind-reading skills to gauge and evoke mating interest, to monitor a mate's commitment, to anticipate a mate's infidelity or defection, and to carry out successful deception. They include self-assessment abilities to evaluate one's mate value, one's future mate value trajectory, and shifts in mate value as a consequence of key life events (e.g. rise or loss of status; gain or loss of a key social ally). They include the other-assessment abilities, such as the capability to monitor the mate value and mate-value trajectories of mates and intra-sexual rivals. And they include the ability to anticipate satellite adaptive problems that follow from deploying particular adaptive solutions. Successful mate guarding, for example, may activate counter-mate guarding adaptations in the mate or in the mate's kin, which create problems that must be solved.

Fitness Signals

Several contemporary human behaviours and cultures associated with mating are derived from ancestral adaptations commonly referred to as 'fitness signals'. Fitness signals evoke essentially 'hard-wired' perceptions that advertise, to the opposite sex, relatively good mating potential, i.e. 'good genes' for passing on to one's children. But these perceptions occur largely without conscious awareness or intent (and of course without literally thinking about genes). In other words, the signals involved (at least in our ancestral past) have virtually always associated accurately with genuinely high relative fitness potential—which is why attraction to them does not need to be accompanied by any awareness of why one is attracted to them. All that matters is that they *were*, for whatever reasons, attractive to ancestors, and this promoted offspring production because the signals were indeed correlated with high genetic fitness (Ridley 1993; Geher and Miller 2008).

Classic fitness signals include perceptions of beauty reflected in qualities such as facial symmetry (especially in women) and wider jaws in men as markers of good health and developmental stability—and sexiness reflected in male preference for a waste-to-hip ratio of around 0.7 that signals fertility (birthing) potential in females. The latter is also signalled by youthful facial appearance in women since their fertility declines sharply with age. Thus, as men age, they commonly prefer partners younger than themselves. But, in contrast, relatively young women commonly prefer older partners because, in our ancestral past (and still today), a man's ability as a 'provider' generally increases with age.

Men are famous for advertising their 'providing' potential (without awareness of its evolutionary roots or potential influence on gene transmission success) by indulging in conspicuous, wasteful accumulation of expensive possessions, and to engage in lavish and excessive leisure. These activities can provide an honest signal of wealth, status, and membership in a privileged class—hence promoting the ability to coerce, attract, and support more mates and their offspring. Showing waste is the only guarantee of truth in advertising (Saad 2007). As a mating tactic therefore, young men have often been anxious to buy things like expensive (usually fast) cars that they can't really afford—and women (especially those middle-aged and still fertile, and/or with young offspring in need of support) have commonly accentuated their waste-to-hip ratios, artificially, using tricks like padding, corsets, and girdles, and have disguised their age using cosmetics. In some cases then the 'advertisement' is not always truthful, but the tricks often work anyway.

For men—especially young ones—displays of athleticism and risk-taking are often also attractive to women because, ancestrally, they have presented as truthful advertisements   of   skill   and   bravery   associated   with   resource-   and protection-providing abilities—i.e. to 'come to the rescue' for a mate and her children (who should be likely therefore to inherit the same male phenotypic qualities that in turn promote success for grandchildren). And these displays have also served to intimidate rival males by signalling a formidable adversary (including in competition for mates). Unfortunately, however, 'run-away' selection has produced a contemporary culture of 'crazy bastards' (Fessler et al. 2014), where less-skilled men are sometimes so eager to participate in the male arena of idiotic risky behaviours (inspired by both social learning and 'whispers' from genetic inheritance), that they suffer dire consequences, or at least embarrassment and loss of pride (Lendrem et al. 2014). Consider 'running of the bulls', skate-boarding stunts, and 'car surfing'; women are rarely seen lining up to participate in these activities, but they may commonly find them sexually attractive. More generally, male displays of the 'dark triad' of traits—narcissism, psychopathy, and Machiavellianism—may be attractive to women because they signal 'mate quality' in terms of confidence and the willingness to take risks (Carter et al. 2014). Marcinkowska et al. (2016) found that women, who were more attracted to men with narcissistic-looking faces, also tended to have more children.

Other fitness signals manifest from some of the human mind's most impressive abilities. Displays of intelligence, skill, and virtue through language, morality, art, and other products of creativity can all be interpreted as courtship tools—costly 'ornaments' evolved to attract and entertain sexual partners. As Miller (2000) put it: 'Our minds are entertaining, intelligent, creative and articulate far beyond the demands of surviving on the plains of Pleistocene Africa'. Men, for example, have been generally more prolific as artists and musicians, advertising intellectual and manual skills to impressionable potential mates. The most accomplished of these today are famous for their large number of sexual partners. This is because the most accomplished of these in our ancestral past signalled a heritable, 'high-quality' mind, and hence a good mating prospect in helping to raise and provide for offspring similarly equipped (through genetic inheritance) with high phenotypic quality. Of course, the adoring female here was not (and didn't need to be) aware that her attraction to the charming suitor was informed by genetic inheritance, nor that its consequence was likely to promote her own gene transmission success.

On Being Male

Throughout our evolution as 'mating machines', natural selection tracked only fitness—not harmonious relationships. In other words, as Richards (2000) explains:

> ... what determines the natures of males and females is not some overall plan. It is just a struggle between genes that happens to have resulted in some good and harmonious elements, but also in a great deal that is not in the least harmonious. ... the puppeteer – natural selection – has zero regard for the feelings of the puppets. And since natural selection is just a matter of the survival of some genes and the disappearance of others, there is not the

slightest reason to expect the sexes to be well designed to make each other happy. This does not mean that it designs them to make each other miserable; it just doesn't design them at all.

Because the probability of gene transmission success for men (but not for women) has always been greater (on average) when the number of matings is maximized, men generally have a stronger sexual impulse (drive) than women (Lippa 2009; Tidwell and Eastwick 2013). Hence, they are more likely to be aroused by visual stimuli (e.g. involving sexualized displays of female bodies), and are more inclined to seek uncommitted sex as often as possible, just for the act of it – e.g. in a 'one-night stand' (Archer 2019). There is nothing new in this observation. Although pre-dating knowledge of Darwinian evolution (by over 2000 years), Aristotle evidently observed: 'Young men have strong passions, and tend to gratify them indiscriminately. Of the bodily desires, it is the sexual by which they are most swayed and in which they show absence of self-control' (Barnes 1984).

Women, in contrast, tend to be more wary and discriminating (Clark and Hatfield 1989). Ancestrally, the biological cost of pregnancy has always been negligible for a man but inordinately high for a woman. 'She had to carry a fetus, then nurse an infant, and then care for a child for years afterward. This is true regardless of whether her pregnancy stems from a one-night stand encounter with charming stranger, or from the thousandth night of conjugal bliss with her loving dedicated husband' (Kenrick and Griskevicius 2013).

These sex differences of course also account for the fact that men are, by far, the main consumers of pornography and the main customers of sex workers. And they are also the main perpetrators of sexual harassment, exploitation, abuse, and assault. It is no coincidence therefore that, historically, men with power to exercise control over secrecy (e.g. priests in certain religious denominations) have traditionally surrounded themselves with two vulnerable groups that have routinely been the victims of sexual abuse at the hands of males—powerless girls and women (e.g. nuns) and powerless young boys (e.g. choirboys and altar boys). Importantly, while these male motivations and behaviours (like cheating, lying, and stealing) are partially informed by the 'whispering genes' of Darwinism, this does not justify them. Ethical and moral codes of conduct are for the public good of societies, while 'whispering genes' act only for their own 'selfish' best interests (copying and transmission success). The latter, therefore, never mean that we cannot, or should not, be held accountable to society for our actions.

Because of differently evolved 'mating intelligence', the potential for ordinary platonic relationships between men and women is also commonly constrained. Historically, the culture of heterosexual interest has presented mostly as follows: women entice; men chase. The potential fitness benefit of pursuing apparent sexual opportunities (multiple matings) for our male ancestors was so great that it vastly out-weighed the cost of occasionally (or even frequently) wasting time pursuing a disinterested woman (a 'false alarm') (Kenrick and Griskevicius 2013). Men, therefore, routinely overestimate the extent to which women are sexually interested in them, and this makes it notoriously difficult for many (perhaps most) men to be 'just friends' with women—who in contrast usually have no difficulty managing such relationships (except when compromised by unwelcome sexual advances from their naïve male friends who end up looking like fools) (Ward 2012). Remarkably, many men it seems are unbothered by this—and commonly even boast to their male friends about their pursuits, regardless of whether they failed.

With inheritance of a hyper-active sex drive and a penchant for idiotic risk-taking behaviour, one might be inclined to interpret from this that biocultural evolution has not been especially 'kind' to men. But there is another, darker side to the statistics of being a male: fear of paternity uncertainty. Prior to the science of genetic testing for paternity, our male ancestors could never have complete (evidence-based) confidence that they were the father of the children they were raising. This mattered for gene transmission success, and so it is not surprising that many male behaviours and cultures have evolved that minimized paternity uncertainty—evident, for example, in the widespread tradition where children are more likely to bear their father's, rather than mother's, last (family) name, and especially where sons are also given the father's first name (Saffa et al. 2021). More troubling, our male predecessors who left the most descendants were those who not only battled successfully with other males over access to mates, but who also exercised rigid control over the sexual activity and fertility of those mates (Low 2005). And the physical strength for success in the former also equipped men for success in the latter—a defining feature of a long history of patriarchal subjugation and cruelty (e.g. Joshi 2006) that men have nothing to be proud of, and which, I would like to think, sickens and horrifies men today.

These controlling behaviours have manifested in several abhorrent traditions instituted by men, many of which, unfortunately, are still with us: forced use of chastity belts (a locking device ostensibly used in medieval times, enabling a man to control when his mate could have sexual intercourse), female genital mutilation (to deny women of sexual pleasure and thus reduce the likelihood of female promiscuity), forced child marriage, honour killings, rape (including especially spousal rape), and other cultures of intimidation, subjugation, and violence against women. Even by just keeping one's mate perpetually pregnant, a man could ensure that she is most of the time infertile (and so any sexual activity with other males would have no effect in threatening his paternity), and also ensure in any event that she will be less attractive to other males because of her enlarged waist-to-hip ratio while pregnant.

These behaviours and cultures are, to a large extent, rooted in a disposition for sexual jealously (distress from the thought of a partner's sexual infidelity) that is famously more acute and conspicuous in men than in women (Buss et al. 1992). But they are also fueled in part, I suggest, by a deeply ingrained 'legacy drive' in humans—motivation to leave 'something of oneself' for the future, rooted in our uniquely human anxiety from awareness of self-impermanence (explored in Chap. 10). In the case of legacy through biological offspring, this implies knowledge of a male contribution (albeit mysterious in its details to our very distant ancestors) as a necessary precursor for human birth (see also Chap. 5). But male control over female fertility is also seen in other animals (e.g. elephant seals and baboons), and so cognitive awareness of what a 'biological father' means (in terms of conception and its relationship to copulation and paternity—knowledge that modern humans have and other animals do not) is unnecessary to account for the early evolution (and persistence) of human male behaviours and cultures that subjugated females. In other words, even in the absence of this cognitive awareness, males who expressed these dominance traits (without needing to know why) undoubtedly left on average more descendants than those who didn't.

Pair-Bonding

It is not surprising that love is a 'beautiful thing'. Humans crave romance and sexual intimacy because it does a pretty good job in promoting gene transmission success. This is an important element of course in the formation of modern dating, marriage, and common-law relationships. But pair-bonding in early foraging humans involved other equally (or probably more) important considerations, including as a basal unit in defining the division of labour needed for efficient group living. Males and females were better suited for different tasks in the business of maintaining provisions for a strong, cooperative social group, including at the smallest scale of individual hearths. Pair-bonding also reduced paternal uncertainty (by instituting male control over female promiscuity), but this also led to increased paternal investment in offspring care and so resulted in fitness benefits for the female partner as well (Ridley 1993).

There have always been potential fitness benefits, however, from secretive adulterous liaisons. For a male, this was a supplemental strategy for further decreasing his uncertainty of paternity. And it allowed a female to keep a partner who is a better bet to support her and her children—including some (or all) of which could be secretly fathered by other men (displaying strong fitness signals—see above) whose genes often would have been superior for producing healthier, stronger children. Men have especially wanted it both ways—being driven to obtain sex with other men's partners, but at the same time driven to deny a partner's opportunity for sex with other men. Male paranoia about being cuckolded has fueled a long history of domestic violence and abuse, including virtual imprisonment of women, and harsh adultery laws against female infidelity (but not for males until relatively recently) in many cultures (Diamond (1992).

According to several accounts (e.g. Roth 2004; Coontz 2005), polygyny has been the dominant model for marriage across human history, at least since the beginnings of wide variation in male wealth and the power that it grants (particularly following the advent of agriculture and empire). The richest men could afford to buy more daughters of other men and so have the most wives—which were essentially owned as 'property'. For men with power then, polygyny always promoted prolificacy and hence evolutionary fitness. Famous extreme examples include Genghis Kahn; 1 in 200 men alive today are estimated to be his direct descendants (Khan 2010). One in five males in northwest Ireland have inherited the Y-chromosome of Niall Noigiallach (Moore et al. 2006)—a powerful Irish king from the fifth century. And a Moroccan emperor, Moulay Ismail the Bloodthirsty (who reigned 1672–1727), is said to have sired 888 children over his lifetime through a harem of 500 women. As Geher and Miller (2008) describe: 'He was a sexist, oppressive, patriarchal psychopath, but evolutionary adaptiveness rarely equals moral virtue (ask any predator or parasite)'.[1]

Ironically (it might seem at first), women could also benefit from a polygynous mating system. In many cases, a woman's potential evolutionary fitness (capacity to leave descendants), as well as her quality of life, would have been greater as the third or fourth wife of a wealthy, high status male, than as the only wife of a poor peasant with no status or earning power (Roth 2004; Lawson et al. 2015). But monogamy has dominated in recent times. As in some other animals, a monogamous culture in humans may have evolved in connection with male fitness benefits from guarding solitary females and reducing infanticide risk from rival males, as well as fitness benefits from greater paternal care when the cost of raising offspring is high (as in human children with their long period of infant dependency needed to grow large brains) (Opie et al. 2013). Monogamy may also have promoted genetic fitness (and social group success) in protecting against sexually transmitted infections—presumably less likely in small hunter-gatherer groups, but increasing in prevalence with development of larger and more crowded agricultural societies (Bauch and McElreath 2016). Limitation (or abolishment) of polygyny can also promote more stable social order by reducing inequality among males. As Wright (1994) put it: 'A polygynous nation, in which large numbers of low-income [sex-starved] men remain mateless, is not the kind of country many of us would want to live in'. Nevertheless, even when monogamy is socially imposed and protected by law, extra-marital affairs—for men, including through mistresses and concubines—have routinely been allowed and even sanctioned in many societies; in other words, '… polygyny tends to lurk stubbornly beneath the surface' (Wright 1994). And the ease and popularity of divorce today means that 'serial monogamy'—multiple partners over time—is probably at least as common as true monogamy.

In recent decades there has been a sharp decline in marriage (Marquardt et al. 2012), coinciding with a growing cultural norm of female empowerment and greater independence (Bolick 2011, 2015). Historical motivations for marriage had nothing to do with love and romance; they involved socio-economic imperatives (Coontz 2005). A spouse's family could be a source of wealth, or might provide advantages linked to social status or alliances. A spouse could also provide offspring for economic purposes—to 'work the family farm', or to look after you in your old age. In general, a woman historically needed to get married in order to secure adequate provisions for herself and her offspring. But the most pressing motivations for marriage were exclusively male: to obtain greater certainty of paternity (because by having a wife, a man could control her sexual activity); to obtain a childcare worker for his offspring; to obtain a housekeeper and cook for himself and his offspring; and to provide a safe, reliable outlet for satisfying the insatiable male sex drive

---

[1] Morality, strategically deployed within one's social group, however, can indeed be adaptive, as we will see in Chap. 8.

(especially when mistresses were in short supply, or if he couldn't afford them). In much of the deep history of patriarchy, husbands were commonly regarded as owners of both their wives and children (signified by having all of them adopt his family name – a tradition that persists today). For many women, the only alternatives to marriage were employment in a grueling, low-paying job, prostitution, or life in a religious order (e.g. convent).

The traditional reasons for marriage however have now essentially vanished in most developed countries, where women are becoming (or are now) largely empowered to take control of their sexuality, their fertility, their livelihood, and their opportunities for societal influence and power—and more likely to be living without a spouse (Trimberger 2005; Davies 2008; Zeidler 2018). These developments are pointing—for better or for worse—to a more marginal role for men in shaping human cultures and politics of the future compared with the past (Tiger 1999; Rosin 2010; Bribiescas 2011). Move over guys.

Parenthood and Family

Biocultural evolutionary roots are also evident in several human affairs associated with parenthood and family. Probably the most obvious is the parental care instinct that humans share widely with many other animals. Most parents say they would risk their own lives to save the lives of their children. There is no doubt that this is 'in our genes'; it is obviously in their best interests.

Many other familial customs and traditions were also in the best interests of the resident genes in our ancestors. One of the most conspicuous is the historically widespread bias in favour of producing and supporting sons. Male offspring, on average, have always had greater potential reproductive value because a son can father many more grand-offspring for his parents over a lifetime than what a daughter can give birth to. We can predict, therefore, that parents of the past (especially in wealthy classes) were on average likely to leave more descendants if their children were all sons than if they were all daughters (Box 7.2). As King Solomon wrote over 3000 years ago: 'Sons's sons are the crown of old men' (Proverbs 17:6, Young's Literal Translation). Ridley 1993 elaborates: '… whenever you look in the historical record, the elites favoured sons more than other classes: farmers in eighteenth century Germany, castes in nineteenth-century India, genealogies in medieval Portugal, wills in modern Canada, and pastoralists in modern Africa. This favouritism took the form of inheritance of land and wealth, but it also took the form of simple care'.

Box 7.2: The Evolution of Preference for Male Offspring

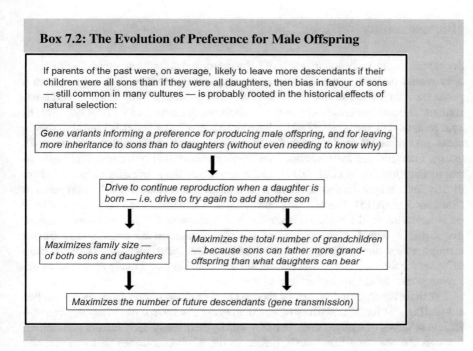

If parents of the past were, on average, likely to leave more descendants if their children were all sons than if they were all daughters, then bias in favour of sons — still common in many cultures — is probably rooted in the historical effects of natural selection:

Gene variants informing a preference for producing male offspring, and for leaving more inheritance to sons than to daughters (without even needing to know why)

Drive to continue reproduction when a daughter is born — i.e. drive to try again to add another son

Maximizes family size — of both sons and daughters

Maximizes the total number of grandchildren — because sons can father more grand-offspring than what daughters can bear

Maximizes the number of future descendants (gene transmission)

However, '… there is a close relationship between social status and the degree to which sons are preferred. … in feudal times, lords favoured their sons, but peasants were more likely to leave possessions to daughters. While their feudal superiors killed or neglected daughters or banished them to convents, peasants left them more possessions' (Ridley (1993). In other words, resource availability is expected to have greater effects on male than on female reproductive success, especially in polygynous societies historically. Here, a male in 'good condition' should out-reproduce his similarly advantaged sister, while the sister should out-reproduce her brother if both are in 'poor condition' because she can marry up, while the poor brother is unlikely to win any mates at all (Trivers and Willard 1973). In many contemporary societies the boy-preferring culture persists, apparently independent of economic or social status (Higginson and Aarssen 2011)—although recent research has detected a signal of its traditional association with wealth (Song 2018).

Evolutionary interpretations have also provided a deeper understanding of several other features of kinship in humans (see review chapters in Salmon and Shackelford 2011)—including sibling rivalry, strategic allocation of maternal effort and paternal effort to particular offspring, variation in parental relationships with genetic offspring versus step-offspring, and variation in grandparental investment as a function of relational uncertainty. As Salmon and Shackelford (2011) write: 'Humans, along with other species, have evolved specialized mechanisms for processing information and motivating behavior relevant to the specific demands of being a mate, father, mother, sibling, child or grandparent'. And in each case, these mechanisms are rooted in consequences measured in terms of ancestral gene transmission success.

Homosexuality

A chapter on the evolutionary roots of behaviours and cultures associated with sexuality would be incomplete without addressing homosexuality. It is unlikely that a preference for same-sex sexual relationships is a product of only learning and environmental/developmental experiences—although these undoubtedly play a significant role, as with most other human behaviours. A role for genes is now abundantly clear, and a recent study concludes that many genes are involved (Ganna et al. 2019). Pedigree and twin studies have clearly shown that homosexuality tends to run in families (Ngun et al. 2011), and a genetic analysis of pairs of gay brothers clearly links sexual orientation in men with particular regions of the human genome (Sanders et al. 2015). The pressing question—a 'Darwinian paradox'—then is: how do we account for the common occurrence of homosexuality in evolutionary terms, given that it would seem to present a severe limitation on evolutionary fitness through one's direct lineage? Several explanations and speculations have been offered (see the very accessible review in Barash 2012)—and many illustrate, as the saying goes: 'things are not always as they seem'.

For male homosexuality, one of the best explanations comes from recent studies suggesting that female relatives of gay men generally have more offspring than the female relatives of straight men. In other words, there are genetic factors transmitted through the maternal line (partly linked to the X-chromosome) that increase the probability of becoming homosexual in males, but they promote higher fecundity in females (Camperio-Ciani et al. 2004; Iemmola and Camperio-Ciani 2009). Hence, the genetic factors that '… influence homosexual orientation in males are not selected against because they increase fecundity in female carriers, thus offering a solution to the Darwinian paradox and an explanation of why natural selection does not progressively eliminate homosexuals' (Iemmola and Camperio-Ciani 2009).

A similar hypothesis (that applies to either male or female homosexuality) is suggested by Zietscha et al. (2008):

> The genes influencing homosexuality have two effects. First, and most obviously, these genes increase the risk for homosexuality, which ostensibly has decreased Darwinian fitness. Countervailing this, however, these same genes appear to increase sex-atypical gender identity, which our results suggest may increase mating success in heterosexuals. This mechanism, called antagonistic pleiotropy, might maintain genes that increase the risk for homosexuality because they increase the number of sex partners in the relatives of homosexuals. … The traits most reliably associated with homosexuality relate to masculinity–femininity; homosexual men tend to be more feminine than heterosexual men, and homosexual women tend to be more masculine than heterosexual women.

In other words, this 'sex atypicality' may be advantageous when expressed in heterosexuals. Do some (perhaps many) females tend to be more attracted to males with certain feminine behavioural traits such as tenderness, considerateness, and kindness? In this study, the results indeed show that '… psychologically masculine females and feminine men are (a) more likely to be nonheterosexual but (b), when heterosexual, have more opposite-sex sexual partners' (Zietscha et al. 2008).

Box 7.3: Bivariate Trait Space Continuum for Sexual Orientation

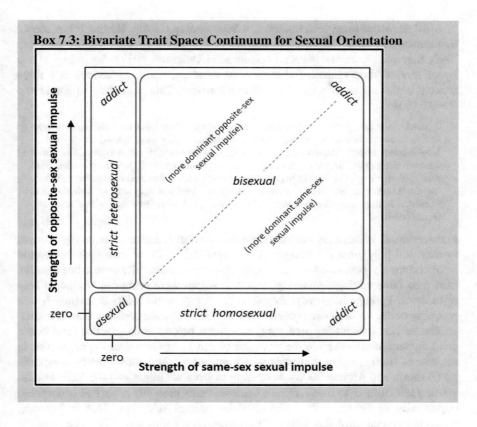

Another, more general hypothesis for homosexuality is that same-sex attraction never really imposed a significant penalty on fitness in our deep ancestral past, *because heterosexual sex was still routinely practiced in spite of it*. There are two very different contexts in which this effect would be expected. The first obtains from just straightforward bisexuality, i.e. where ancestral sex lives commonly involved a mix of both same-sex and heterosexual activity but in varying proportions (informed in part by genotypic variability) ranging from bisexual with a more dominant opposite-sex attraction, to bisexual with a more dominant same-sex attraction (Box 7.3). But random mating within this mix would also have produced genetic variants that informed strictly homosexual as well as strictly heterosexual orientations. Importantly, only the latter is strongly favoured by natural selection. Nevertheless, despite being strongly disfavoured by natural selection, strict homosexuality would have persisted in low frequency simply because of genetic factors informing same-sex attraction that were inherited from bisexual ancestors. Under this hypothesis, homosexuality is not really a Darwinian paradox at all; instead it, along with asexuality, were just periodic maladaptive genotypic by-products of ancestral gene transmission.

Most ancestral bisexuals, however, were probably female. Research has shown that men are generally attracted to one sex or the other, whereas women are more likely than men to have a bisexual orientation (Kanazawa 2017). According to one hypothesis, a 'fluid sexuality' that enabled same-sex sexual behaviour in women made it easier for women to raise children together. This, according to Kuhle and Radtke (2013):

> ... would have been particularly beneficial to ancestral women when their male mates were unable to adequately care, protect, and provide because they were injured, away on prolonged hunts, or preoccupied finding, courting, and mating with other women. The latter scenario was particularly likely to occur within polygynous mating systems. If so, it is possible that men's relative lack of aversion to a female mate's homosexual, rather than heterosexual, affair ... and men's common fantasy of simultaneously mating with multiple women ... is an outgrowth of a male psychology designed to promote their mates' same-sex sexual behavior.

Female sexual fluidity may have also minimized conflict and tension among cowives in ancestral polygynous marriages (Kanazawa 2017). These consequences would have commonly benefitted the offspring of not just bisexual women, but also the men who fathered these offspring, many of whom were often not around to help raise them. Perhaps ironically, therefore, for many of our male ancestors, it was probably better for their own genetic fitness, if their female partners were bisexual.

The second context for predicting successful heterosexual practice (and hence gene transmission)—despite the presence of same sex attraction—applies even in the case of strong *preference* for same-sex encounters, including strict homosexuality (Aarssen and Altman 2006). According to what we might call the 'failed disfavouring selection' hypothesis, female homosexuality probably never had widespread opportunity to be strongly disfavoured by natural selection. This is because, throughout much (probably most) of human history, males were to a large extent in control of the sexual activity and fertility of females (as discussed earlier in this chapter). Many or most women, therefore, were essentially forced—by patriarchal subjugation, socio-cultural expectations, and/or religious imperatives—to mate with men and bear their (frequently many) offspring, *regardless of their sexual orientation*. [The same would also have been true regardless of the intensity of female sexual impulse (drive)]. Accordingly, so-called gay genes in many of our female ancestors—including those that might have been inherited by their sons as well as their daughters—were never significantly limited in their transmission success to future generations. And so in this context, the prevalence of homosexuality today is (again) not really a Darwinian paradox at all.

Importantly however, women more than ever are now in control of both their fertility and their sex lives, and their empowerment for this and other basic human rights continues to grow rapidly on a global scale. For our predecessors, choosing successfully to be a practicing homosexual meant zero gene transmission through direct lineage. But today this is not necessarily the case, with reproductive technologies for sperm banks and in vitro fertilization (and perhaps, in the future, human cloning). But unless the latter become widely practiced as a routine cultural norm, the widespread mating and reproductive freedom for women today means that selection against an exclusively or predominantly lesbian orientation may soon be ramping up (Aarssen and Altman 2006). If female bisexuality remains alive and well, however, so also would some homosexuality (because of genetic inheritance from bisexual maternal ancestors).

The above scenario is a good example of how some traits (like homosexuality, and the generally weaker sexual impulse in females compared with males), rather than a consequence of being favoured by natural selection in the ancestral past, may instead be a consequence of *not having been disfavoured by natural selection*. In Chaps. 11 and 12, we will return to this hypothesis in our examination of more recent Darwinian paradoxes: the 'childfree' culture, the 'freemale' culture (introduced in the present chapter), and the 'sex recession culture'—and their implications for addressing the critical challenges that we face as global population size continues to rise.

References

Aarssen LW, Altman S (2006) Explaining below-replacement fertility and increasing childlessness in wealthy countries: legacy drive and the "transmission competition" hypothesis. Evol Psychol 4:290–302

Anders KM, Olmstead SB (2019) "Stepping out of my sexual comfort zone": comparing the sexual possible selves and strategies of college-attending and non-college emerging adults. Arch Sex Behav 48:1877–1891

Archer J (2019) The reality and evolutionary significance of human psychological sex differences. Biol Rev 94:1381–1415

Barash DP (2012) Homo mysterious: evolutionary puzzles of human nature. Oxford University Press, Oxford

Barnes J (ed) (1984) The complete works of Aristotle: the revised Oxford translation. Princeton University Press, Princeton

Bauch C, McElreath R (2016) Disease dynamics and costly punishment can foster socially imposed monogamy. Nat Commun 7:11219. https://doi.org/10.1038/ncomms11219

Bolick K (2011) All the single ladies. Atlantic Magazine, Nov 2011. http://www.theatlantic.com/magazine/archive/2011/11/all-the-single-ladies/308654/?single_page=true

Bolick K (2015) Spinster: making a life of one's own. Crown Publishers, New York

Bribiescas RG (2011) An evolutionary and life history perspective on the role of males on human futures. Futures 32:729–739

Buss DM (2008) Forward: the future of mating intelligence. In: Geher G, Miller G (eds) Mating intelligence: sex, relationships and the mind's reproductive system. Lawrence Erlbaum Associates, New York

Buss DM, Larsen RJ, Westen D, Semmelroth J (1992) Sex differences in jealousy: evolution, physiology, and psychology. Psychol Sci 3:251–256

Camperio-Ciani A, Corna F, Capiluppi C (2004) Evidence for maternally inherited factors favouring male homosexuality and promoting female fecundity. Proc R Soc London Ser B 271:2217–2221

Carter GL, Campbell AC, Muncer S (2014) The Dark Triad personality: attractiveness to women. Personal Individ Differ 56:57–61

Clark RD, Hatfield E (1989) Gender differences in receptivity to sexual offers. J Psychol Hum Sex 2:39–55

Coontz S (2005) Marriage – a history: from obedience to intimacy, or how love conquered marriage. Viking Press, New York

Cox A, Fisher M (2009) The Texas billionaire's pregnant bride: an evolutionary interpretation of romance fiction titles. J Soc Evol Cult Psychol 3:386–401

Davies C (2008) Single and happy: it's the freemales. The Guardian, 13 Apr 2008. https://www.theguardian.com/lifeandstyle/2008/apr/13/women.familyandrelationships3

Diamond J (1992) The third chimpanzee: the evolution and future of the human animal. Harper, New York

Fessler DMT, Tiokhin LB, Holbrook C, Gervais MM, Snyder JK (2014) Foundations of the crazy bastard hypothesis: nonviolent physical risk-taking enhances conceptualized formidability. Evol Hum Behav 35:26–33

Ganna A et al (2019) Large-scale GWAS reveals insights into the genetic architecture of same-sex sexual behavior. Science 365(6456):eaat7693. https://doi.org/10.1126/science.aat7693

Geher G, Miller G (eds) (2008) Mating intelligence: sex, relationships and the mind's reproductive system. Lawrence Erlbaum Associates, New York

Higginson MT, Aarssen LW (2011) Gender bias in offspring preference: sons still a higher priority, but only in men — women prefer daughters. Open Anthropol J 4:60–65

Hobbs DR, Gallup GG (2011) Songs as a medium for embedded reproductive messages. Evol Psychol 9:390–416

Iemmola F, Camperio-Ciani A (2009) New evidence of genetic factors influencing sexual orientation in men: female fecundity increase in the maternal line. Arch Sex Behav 38:393–399

Joshi ST (ed) (2006) In her place: a documentary history of prejudice against women. Prometheus Books, Amherst

Kanazawa S (2017) Possible evolutionary origins of human female sexual fluidity. Biol Rev 92:1251–1274

Kenrick DT, Griskevicius V (2013) The rational animal: how evolution made us smarter than we think. Basic Books, New York

Khan R (2010) 1 in 200 men direct descendants of Genghis Khan. https://www.discovermagazine.com/the-sciences/1-in-200-men-direct-descendants-of-genghis-khan

Kuhle BX, Radtke S (2013) Born both ways: the alloparenting hypothesis for sexual fluidity in women. Evol Psychol 11:304–323

Lawson DW, James S, Ngadaya E, Ngowi B, Mfinanga SGM, Borgerhoff Mulder M (2015) No evidence that polygynous marriage is a harmful cultural practice in northern Tanzania. Proc Natl Acad Sci 112:13827–13832

Lendrem B, Lendrem D, Gray A, Isaacs J (2014) The Darwin awards: sex differences in idiotic behaviour. BMJ 349:g7094. https://doi.org/10.1136/bmj.g7094

Lippa RA (2009) Sex differences in sex drive, sociosexuality, and height across 53 nations: testing evolutionary and social structural theories. Arch Sex Behav 38:631–651

Low BS (2005) Women's lives there, here, then, now: a review of women's ecological and demographic constraints cross-culturally. Evol Hum Behav 26:64–87

Marcinkowska UM, Lyons MT, Helle S (2016) Women's reproductive success and the preference for Dark Triad in men's faces. Evol Hum Behav 37:287–292

Marquardt E, Blankenhorn D, Lerman RI, Malone-Colón L, Wilcox WP (2012) The president's Marriage Agenda for the Forgotten Sixty Percent. The State of our Unions. National Marriage Project and Institute for American Values, Charlottesville

Miller G (2000) The mating mind: how sexual choice influenced the evolution of human behaviour. Anchor Books, New York

Moore LT, McEvoy B, Cape E, Simms K, Bradley DG (2006) A Y-chromosome signature of hegemony in gaelic Ireland. Am J Hum Genet 78:334–338

Ngun TC, Ghahramani N, Sánchez FJ, Bocklandt S, Vilain E (2011) The genetics of sex differences in brain and behavior. Front Neuroendocrinol 32:227–246

Opie C, Atkinson QD, Dunbar RIM, Shultz S (2013) Male infanticide leads to social monogamy in primates. Proc Natl Acad Sci 110:13328–13332

Richards JD (2000) Human nature after Darwin: a philosophical introduction. Routledge, London

Ridley M (1993) The red queen: sex and the evolution of human nature. HarperCollins, New York

Rosin H (2010) The end of men. Atlantic Magazine, July/Aug 2010. http://www.theatlantic.com/magazine/archive/2010/07/the-end-of-men/308135/

Roth EA (2004) Culture, biology and anthropological demography. Cambridge University Press, Cambridge

Saad G (2007) The evolutionary bases of consumption. Lawrence Elbaum Associates, London

Saffa G, Štěrbová Z, Prokop P (2021) Parental investment is biased toward children named for their fathers. Hum Nat 32:387–405

Salmon C, Shackelford TK (2011) The Oxford handbook of evolutionary family psychology. Oxford University Press, New York

Sanders AR, Martin ER, Beecham GW, Guo S, Dawood K, Rieger G, Badner JA, Gershon ES, Krishnappa RS, Kolundzija AB, Duan J, Gejman PV, Bailey JM (2015) Genome-wide scan demonstrates significant linkage for male sexual orientation. Psychol Med 45:1379–1388

Song S (2018) Spending patterns of Chinese parents on children's backpacks support the Trivers-Willard hypothesis: results based on transaction data from China's largest online retailer. Evol Hum Behav 39:336–342

Tidwell ND, Eastwick PW (2013) Sex differences in succumbing to sexual temptations: a function of impulse or control? Personal Soc Psychol Bull 39:1620–1633

Tiger L (1999) The Decline of Males. St. Martin's Griffin, New York

Trimberger EK (2005) The new single woman. Beacon Press, Boston

Trivers RL, Willard DE (1973) Natural selection of parental ability to vary the sex ratio of offspring. Science 179:90–92

Ward AF (2012) Men and women can't be just friends. http://www.scientificamerican.com/article/men-and-women-cant-be-just-friends/

Wright R (1994) The moral animal. Vintage Books, New York

Zeidler M (2018) Single women increasingly pursuing parenthood on their own. CBC News, 23 June 2018. https://www.cbc.ca/news/canada/british-columbia/single-women-increasingly-pursuing-parenthood-on-their-own-1.4717327

Zietscha BP, Morley KI, Shekar SN, Verweij KJH, Keller MC, Macgregor S, Wright MJ, Bailey JM, Martin NG (2008) Genetic factors predisposing to homosexuality may increase mating success in heterosexuals. Evol Hum Behav 29:424–433

Chapter 8
Staying Alive

As many more individuals of each species are born than can possibly survive; and as, consequently, there is a frequently recurring struggle for existence, it follows that any being, if it vary however slightly in any manner profitable to itself, under the complex and sometimes varying conditions of life, will have a better chance of surviving, and thus be naturally selected. (Darwin 1859)

Maspero G. (1897). *The Dawn of Civilization – Egypt and Chaldaea*/Public Domain (https://www. flickr.com/photos/internetarchivebookimages/14783494613/sizes/l/)

Successful gene transmission requires sex, but sex requires staying alive long enough to have some of it. For humans this requires, at a minimum, growing and surviving through more than a decade of sexual immaturity. In other words, the 'mating machine' depends on a functioning 'survival machine'. And more survival for our ancestors (as in all animals) generally meant more sex. Evolution, therefore (as in all animals), has given us a survival instinct. In humans this involves motivations and behaviours for garnering a wide range of particular resources (e.g. physiological necessities, shelter, social life, and alliances), and for avoiding a wide range of health and mortality risks (injury and disease). In this chapter, we explore how many of our likes and dislikes today, our joys and our fears (and cultures that represent them), were shaped by the evolution of a 'Survival Drive' in Pleistocene Africa (Curtis 2013; Orians 2014). As Kenrick and Griskevicius (2013) put it, 'The humans who became our ancestors were those who protected themselves from enemies and

L. Aarssen, *What We Are: The Evolutionary Roots of Our Future*,
https://doi.org/10.1007/978-3-031-05879-0_8

predators, avoided infection and disease, got along with other people in their tribe, and gained the respect of their fellow tribe members'.

Motivation theories have inspired many decades of research in the behavioural sciences (Deci and Ryan 2000; Shah and Gardner 2008; Cosmides and Tooby 2013). In one of the early influential models, Maslow's (1943) 'pyramid of needs' defines several more or less universal features of human nature in terms of a hierarchical series of motivations. Thus the most basal categories are 'immediate physiological needs' followed by 'safety', and higher order needs are distinguished by 'love (affection, belongingness)' and 'esteem (respect)', with 'self-actualization' occupying the apex of the pyramid. The pyramidal architecture then serves to represent that higher order needs are normally not attained unless more basal needs are met first and that these commonly manifest along a developmental trajectory with advancing age.

An important update of the Maslow pyramid was proposed by Kenrick et al. (2010) to give it a more explicitly Darwinian framework, firmly grounded in modern evolutionary theory—i.e. where motivations are linked to their presumed/probable functions as adaptive cognitive domains in rewarding the reproductive success of ancestors. This approach has parallels in the more recent 'Selfish Goal' model of Huang and Bargh (2014) and in an earlier account of self-determination theory (Deci and Ryan 2000), which also regards '…psychological needs as universal aspects of human nature…' and which '… fits broadly in an adaptationalist perspective that emphasizes how our common evolutionary heritage produces such regularity' (p. 252); and see also Bernard et al. (2005). All of this echoes Darwin's (1872) prescient interpretation of affective states in humans (Keltner and Gross 1999).

In the Kenrick et al. (2010) pyramid 'renovation' (Box 8.1) (see also Kenrick and Griskevicius 2013), Maslow's lower and mid-level needs are essentially retained (with some revised labelling), but the major and significant distinction is the replacement of the pyramid apex by three goals drawn from evolutionary life history theory (discussed in Chap. 7): mate acquisition, mate retention, and parenting. In this renovation, Maslow's 'self-actualization' is not regarded as a functional need, and is considered instead to be '… largely subsumed within status (esteem) and mating-related motives' (Kenrick et al. 2010, p. 239). Others, however, have called for retaining greater emphasis on components of self-actualization and meaning within the needs/goals/motivations framework (Heine et al. 2006; Dweck and Grant 2008; Kesebir et al. 2010; Peterson and Park 2010), and this chimes with an emerging field of research in existential psychology (Greenberg et al. 2008; Vess et al. 2009; Schnell 2012; Shaver and Mikulincer 2012; Batthyany and Russo-Netzer 2014; van Bruggen et al. 2015). These considerations are developed further in Chaps. 9 and 10.

Box 8.1: Depiction of the Fundamental Needs of Humans by Kenrick et al. (2010). See Text

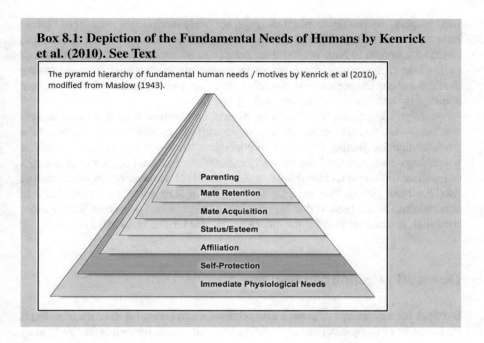

The pyramid hierarchy of fundamental human needs / motives by Kenrick et al (2010), modified from Maslow (1943).

Parenting
Mate Retention
Mate Acquisition
Status/Esteem
Affiliation
Self-Protection
Immediate Physiological Needs

Physiological/Resource Motivations

Negative emotions like fear, anger, and disgust are adaptive of course because they trigger responses that routinely promoted ancestral survival. Hunger and thirst are obviously hard-wired—maintained across generations by adaptive consequences. A baby doesn't need to learn to desire and seek nourishment. What is less obvious are particular cravings and repulsions that have been shaped by natural selection. For example, foods with high contents of sugar, fat, and/or salt are widely abundant today from food retailers, and hard-to-resist cravings for these foods are a common affliction, with health risks for many people, especially the sedentary. But such cravings would have been adaptive for our more physically active distant ancestors, because these foods provided an important source of calories and essential nutrients that were often in short supply, or available only occasionally from the natural environment. And so there was little risk of overindulgence; obesity would have been practically unknown in early foraging humans.

Certain other potential foods trigger universal disgust (e.g. from the smell of spoiled meat) or symptoms of illness if consumed—behaviours and consequences that obviously promoted ancestral gene transmission success. Not surprisingly, these have in some cases informed characteristic features of culture—e.g. traditions of taboo against certain foods, especially meats, that routinely contain pathogens (that are more active at higher temperatures). Hence, some cultures, associated with food preparation and favoured flavourings, are fairly obviously a product of biocultural evolution (Chap. 6). For example, many spices such as chilies, garlic, and onions are known to have healthful antimicrobial properties, and the proportion of indigenous recipes using these spices increases across countries with increasing average annual temperature—coinciding with the greater historical risk of meat spoilage by bacteria (Sherman and Billing 1999).

An interesting example of resource motivation involves what is known as 'loss aversion'—i.e. people are commonly more moved emotionally by the loss of a resource than by gaining a resource of equivalent value—interpreted as an evolved psychology rooted in the frequent and longstanding ancestral costs and challenges of resource limitation and foraging (e.g. Li et al. 2012). In other words, once enough food has been gathered for the daily feeding, losing it represents a potential cost to survival (and hence fitness) that is greater than the potential gain from finding more food than is really needed for the day (Kenrick and Griskevicius 2013).

Defence/Protection Motivations

Survival for any animal of course also requires motivations for defence (e.g. fight, flight, or freeze) against harm, providing obvious needs for safety in the face of danger. Fear of snakes, for example, is virtually universal in humans (LoBue and DeLoache 2010); wariness around them was generally in the best interests of our ancestor's genes, even if much of the time, the danger was negligible. For the same reason, humans are impulsively wary of heights. According to 'Evolved Navigation Theory' (ENT), natural selection has shaped height perception in response to the navigational outcome of falling. In one study, participants perceived greater height from the top than from the bottom of a vertical surface—consistent with the predictions of ENT (because descent results in falls more often than does ascent) (Jackson and Cormack 2007).

Research has shown that several adaptive human behaviours involving wariness and caution (being 'on guard') are triggered instinctively when humans are primed with the information that signals potential risk of harm. Kenrick and Griskevicius (2013) summarize:

> ... your paranoid inner self can be primed not only by real or perceived physical danger but also by angry expressions on the faces of strangers, thoughts about members of other races and religions, scary movies, or the local evening news broadcast which often starts with a gruesome crime ... This natural proclivity toward vigilance and paranoia inclines people to invest resources to avoid weakness or vulnerability, while placing high value on attaining strength or associating themselves with a powerful numerical majority.

Other kinds of evolved self-protection motivations involve mechanisms for avoiding contact with infectious pathogens/disease. These can be activated by '...things like the sound of other people sneezing and coughing, the sight of skin lesions and foul smells'. And (for some) '... by thinking about people from exotic faraway places such as Sri Lanka and Ethiopia as opposed to Hawaii or London'

(Kenrick and Griskevicius 2013). As Smith (2007) put it, humans have evolved an 'anti-parasite module':

> When human beings lived in small isolated groups, encounters with strangers were potentially threatening because you might not have acquired a resistance to the germs carried by the outsider." Hence, "... vulnerability to disease may underpin our hostility toward strangers. ... The robustness of disgust reactions and their universality suggest they are deeply rooted in human nature, and that our ancestors evolved a mental module specifically for dealing with the risks of parasitic infection.

Research has shown that women in early pregnancy—when the foetus is most vulnerable to complications from a diseased mother—can become especially afraid of foreigners (Navarrete et al. 2007). Nettle (2009) documents several examples of between-population variation in social structure and norms that correlate strikingly with the degree of local pathogen prevalence—illustrating how cultural variation can be plausibly interpreted in terms of probable effects of biological adaptation/ evolution.

Affiliation/Alliance Motivations

Our ancestors of course needed to mate successfully, but if they were otherwise anti-social, preferring to fend entirely for themselves as solitary individuals, they left few descendants. Humans are more or less obligately social, and several behaviours and cultural norms are born out of this strong social instinct—an evolved psychology prescribing a need to 'belong' and to form cooperative alliances, affiliations, and friendships with other in-group members. Both religion (Chap. 10) and alcohol production and consumption (Slingerland 2021) emerged as two of the most profound products of this biocultural evolution (Slingerland 2021), each playing a pivotal role in the progress march of civilization (Chap. 5).

This strong social instinct explains why we feel loneliness (Sarner 2017), why we dance (Tarr 2017), why we commonly greet, and make agreements with a handshake (Al-Shamahi 2021), and why we laugh, cry, and smile—sometimes while faking it (Graziano 2014; Knight 2019; Bergson 2020; Danvers 2021). It explains our sense of self-esteem and our tendency to be insecure (providing a monitor of our level of acceptance by others) (e.g. Hendriksen 2018), as well as our emotions of empathy, guilt, and shame and our cultures of ethics, virtues, and gossip. Several extensive accounts provide reviews on these themes (e.g. Ridley 1998; Joyce 2006; Henrich and Henrich 2007; Tomasello 2009; Bowles and Gintis 2011; Boehm 2012; Bloom 2013; Hare and Woods 2020). Attraction to gossip, for example, is virtually universal in humans. From McAndrew (2008):

> Our prehistoric ancestors had to cooperate with so-called in-group members for success against out-groups, but they also had to recognize that these same in-group members were their main competitors when it came to dividing limited resources. Living under such conditions, our ancestors faced a number of consistent adaptive problems such as remembering who was a reliable exchange partner and who was a cheater, knowing who would be a

reproductively valuable mate, and figuring out how to successfully manage friendships, alliances and family relationships. … The social intelligence needed for success in this environment required an ability to predict and influence the behavior of others, and an intense interest in the private dealings of other people would have been handy indeed and would have been strongly favored by natural selection. In short, people who were fascinated with the lives of others were simply more successful than those who were not, and it is the genes of those individuals that have come down to us through the ages.

This also accounts for why humans are routinely attracted to sensational news stories—i.e. because they trigger an evolved tendency to attend to categories of information that increased ancestral gene transmission (Davis and McLeod 2003).

For the same reason, most humans have an intrinsic moral sense. Helping others in need is universally defined as 'good'—a personal virtue to admire and aspire to—reinforced by both 'whispering genes' and social learning. Moral standards, after all, are 'pro-social'; they support the 'common good', and are the foundation of many cultural codes of conduct, and secular laws (e.g. against stealing and murder). Social groups (and hence their resident membership) that espoused moral standards in our ancestral past were generally more prosperous than those that did not. Accordingly, our moral instincts, complemented by moral codes of conduct, are products of a long history of biocultural evolution, spanning several hundred thousand years—starting from 'individual intentionality' (self-interest), progressing to 'joint intentionality' (collaborative foraging), and arriving, about 100,000 years ago, at 'collective intentionality' (sets of practices, goals, conventions, and institutions) that distinguished a tribe (Tomasello 2018).

Importantly, there is nothing intrinsically selfless or altruistic about one's motivation to be kind, friendly, and helpful to others; our predecessors who had these dispositions (and especially those inclined to deploy them strategically) left more descendants than those who did (were) not. Accordingly, religion is not needed in order to have an objective basis for moral standards. Morality is in our genes, and it is in our religions only because it is in our genes. [We return to the topic of religion in Chap. 10.]

Helpful behaviour evolves most readily through 'kin-selection', i.e. when recipients are kin, since (compared with non-kin) they share more genes with the helper. Because of this help, the shared genes obtain greater fitness (transmission success), which is to say that the helper obtains greater *inclusive fitness*. Some of this kin-helping behaviour undoubtedly 'spills over' to benefit non-kin from time-to-time (without any significant cost to the benefactor) simply because they happen to be around, nearby, and are hence fortuitously on the 'receiving end'—and so this might easily be mistaken for altruism (Aarssen 2013). But helping behaviour can also evolve through mutualism and reciprocal exchanges between non-kin, in-group members; e.g. pre-historic hunting of large animals, which provided food benefits for individuals, required group cooperation. Cartwright (2008) expounds:

Morality is the name we give to those emotions and inclinations that steer us away from the temptation to cheat and reap immediate selfish rewards towards cooperative behaviour that helps us to reap the benefits of mutualism …. Emotions such as sympathy, empathy and compassion enable us to experience the perspective of others and bind communities together. The crucial point is that whereas the original evolution of these tendencies took place according to the cold and ruthless logic of natural selection—they were selected for if they gave an advantage to the genes of those possessing them—we are now left enriched by these feelings[1] that spill over into a whole range of contexts, enabling us to feel empathy with and pity for the suffering even when we 'know' cognitively that any help we give is unlikely to be returned even by indirect means. An analogy for this might be sexual attraction. Experiencing sexual urges for someone highly attractive was once an efficient way ensuring that your genes paired up with a high quality complementary set. But now we can experience the urge and enjoy the behaviour it directs even when we are told that the other person is infertile (through contraception, for example).

When helping the needy and other acts of kindness are visible to others, one's reputation in the community is enhanced—e.g. by advertising (honestly or not) a good and trustworthy person—thus earning social privileges, and/or signalling to potential mates that one is likely to be a caring and reliable partner with whom to have and raise children (Aarssen 2013). A similar boost to reputation—through signalling (truthfully or not) that one can be trusted ('virtue signalling')—might also be earned by 'moral outrage/grandstanding', i.e. publicly condemning bad/immoral behaviour in others (Jordan et al. 2016; Levy 2019). For our ancestors, helping others may also have been a subconscious strategy to help ensure future reciprocity, i.e. in the event that one day, charity may be similarly needed for oneself in return. Conspicuous philanthropic giving behaviours by the wealthy can also serve to signal social status, thus conferring certain privileges and attractiveness to potential mates, as well as satisfying 'Legacy Drive' (discussed in Chap. 10). It is doubtful, therefore, that *true* altruism—i.e. sacrificing some of one's genetic fitness in order to benefit unrelated others—really exists. At least there is no unequivocal evidence that it exists, except perhaps rarely, as anomalies. As Wilson (1978) put it, 'We sanctify true altruism in order to reward it, and thus to make it less than true, and by that means to promote its recurrence in others'.

Moral Obligations

There is an important, broader question that underlies these themes on morality: *Does one have greater personal obligation to the welfare of some individuals over others?* The answer is clearly yes for most people, at least in practice, and evolutionary theory explains why—based on genetic relatedness (Box 8.2). Most people of course care immensely about their own personal welfare because we are 100% related to ourselves—ancient survival instinct in action. But one's sense of obligation to others generally falls off with decreasing genetic relatedness. Hence, most people would save the life of a sibling if they could, but perhaps not if it meant risking one's own life; after all, we share only about 50% of our genes with a sibling.

---

[1] Importantly, these enrichments and pleasurable feelings from things like empathy and compassion may also be deployed as components of 'Leisure Drive'—discussed in Chap. 9.

We might be more willing therefore to risk our own life if the effort could save the lives of two siblings (thus doubling the number of shared, savable gene copies). Identical twins reportedly feel a greater obligation to each other than to other siblings—again not surprisingly; they are 100% genetically related to each other (and only about 50% related to other siblings).

Most of us would also be willing to risk (even sacrifice) our own life if it was necessary to save the life of one of our children, or even a grandchild. This might seem at first to contradict the genetic relatedness criterion, since we share (respectively) only 50% and only about 25% of our genes with our children and grandchildren. But because our own individual welfare will inevitably come to an end (with mortality), we have a deeply ingrained parental care instinct. This was inherited from ancestors who didn't know (and didn't need to) that the reason for it was because offspring and grand-offspring are normally the only 'vehicles' that can propel direct copies of one's resident genes into future generations. It is less obvious however, as to why humans are routinely more likely to risk their lives for their children than for their siblings, given that they share 50% of their genes with each of them (Aarssen 2009).

Box 8.2: Does One Have Greater Personal Obligation to the Welfare of Some Individuals Over Others?

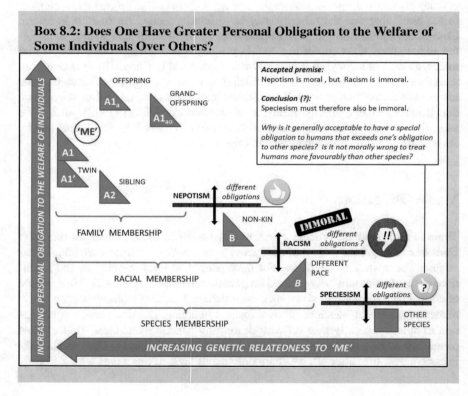

This greater sense of obligation to kin than to non-kin—i.e. bias based on family membership, or nepotism—is cross-cultural, and generally accepted as moral, without debate; it is 'hard-wired' in the 'selfish genes' of humans. But a kind of moral dilemma unfolds as we move further down the genetic-relatedness scale, where we encounter additional levels of in-group bias—most notably based on racial membership (racism), and further down based on species membership (speciesism) (Box 8.2). These are anciently evolved components of human psychology—evident even in babies as young as 6–9 months (Kelly et al. 2005)—because in-group bias (including other forms based on nationalism, xenophobia, enthnocentrism, religion, and other cultural worldviews) (Brown 2004; Kteily et al. 2015) generally rewarded gene transmission success in our deep ancestral past (Douglas 2017; van Prooijen 2019). However, they have commonly received variable ratings in terms of cultural codes of conduct, and the dilemma then lies in defining (or whether it is even possible to define) an objective basis for moral standards.

On the one hand, the bounds of morality are shaped to a large extent by social learning within local cultures. In other words, societies have always decided for themselves what is unacceptable, immoral, and unlawful for the best interests of the 'public good'—and some groups have made better decisions, for their own welfare, than others (consider the success of Nazism). Darwinism can never be used to justify behaviours that violate or compromise human rights or public safety, such as sexual harassment/abuse. Moral and social responsibility always trump genetic bequeathal (see more on this in Chap. 12), and this frequently involves an obligation to say 'no' to some of our 'whispering genes'. At the same time, however (thankfully), humans also have 'whispering genes' that have given us a moral instinct. This is because what is good for the best interests (prosperity) of the social group— e.g. in the case of certain agreed-upon moral codes of conduct—is normally (as emphasized in Chap. 2) contingent on this also being in the best interests of gene transmission success for at least most of the resident individuals. In other words, (as emphasized in Chap. 6), traits influenced by genes have commonly shaped the cultures that people like—because they were (commonly) cultures that rewarded ancestral gene transmission success.

For example, mass murder within a society is of course defined by cultures virtually everywhere as ethically immoral and illegal, and our predecessors inclined towards this behaviour normally also paid a heavy fitness penalty, at least since the origin of social law and order. Our evolution, therefore, may have involved a kind of 'self-domestication' — i.e. killing off the most aggressive and violent males, thus eliminating them from the local gene pool (Wrangham 2019). Accordingly, helpful, cooperative citizens are much more common today than bullies and mass murderers, not just because of effects from learning prosocial cultural norms, but also because of genetic inheritance. Humans have evolved a 'moral instinct' then, not in order to deliver us morality per se—only fitness. Importantly though, there is a catch, as examined below: 'Although morality is commonly defined as involving justice for all people, or consistency in the social treatment of all humans, it may have arisen for immoral reasons, as a force leading to cohesiveness within human groups but specifically excluding and directed against other human groups with different interests' (Alexander 1985). As Sterelny (2012) put it: 'Our prosocial dispositions were forged in the crucible of war'.

A comparison of racism versus speciesism particularly shows the confounds of defining moral standards (Box 8.2). Both—like nepotism—have been regarded as largely consistent with moral codes of conduct throughout much of human history; they were part of our evolution in 'becoming human' (Chap. 3), and in the 'March of Progress' (Chap. 5). But in most cultures today, racism is of course widely regarded as immoral, because it is never in the best interests of modern society. It undermines the 'common good'. And yet, racism often still rears its ugly head. In contrast, speciesism today is still largely regarded as moral (or at least not immoral) by the general public, and the philosopher might reasonably ask why—i.e. along a continuum of genetic relatedness (Box 8.2), why should an immoral in-group bias (racism) be sandwiched between two that are traditionally considered moral (nepotism and speciesism) (e.g. see Lawlor 2012)?

Clearly the harvesting and exploitation of animals for human consumption has been a major boost to human prosperity and genetic fitness ever since our distant ancestors became hunters. Speciesism, however, has come under increasing attack in recent years over concerns about the ethical treatment of animals used in modern medical research and food production. And a growing number of people now regard speciesism to be just as immoral as racism (e.g. http://speciesismthemovie.com/). It will be interesting to watch how this movement unfolds, and to consider whether it has evolutionary roots. For example, might this involve a form of virtue signalling or moral grandstanding, connected with motivation for social status or mate attraction? Or has our empathic instinct evolved to become so acute that the emotion involved (modulated by social learning) has started recently to spill over somewhat towards the plight of individuals belonging to other species? Could this be connected with our uniquely human mortality salience and self-impermanence anxiety (explored in greater detail in the next two chapters)? As Cave (2014) put it:

> *This horror at the death of other creatures is intimately bound up with horror at the prospect of one's own demise. Flies come and go in countless masses, mostly beyond my sight and care. But when something happens that causes me to empathise, to become the fly, then its death becomes terrible. As the poet William Blake realised when he, too, carelessly squashed an insect:*
>
> *Am not I*
> *A fly like thee?*
> *Or art not thou*
> *A man like me?*

In stark contrast, however, as we examine below, humans have always been frightfully willing and ready, in certain circumstances, to impose death on certain other fellow humans.

Us Versus Them (Again)

Throughout most of recorded history (and probably pre-history as well), humans have been to a large extent engaged in conflict with one out-group or another. And when not at war, our ancestors were largely pre-occupied with preparing for it, or recovering from it. Recall from Chap. 5, that these challenges defined probably the single most compelling motivation for technological advances and cultural evolution within societies. As Barash (2008) put it:

> The history of civilization is, in large part, one of ever-greater efficiency in killing: with increasing ease, at longer distances, and in larger numbers. Just consider the 'progression' from club, knife, and spear to bow and arrow, musket, rifle, cannon, machine gun, battleship, bomber, and nuclear-tipped ICBM.

It is a sobering thought that we are probably all descendants of the instigators and winners of wars at one time or another in our ancestral past. Losing wars was not good for genetic fitness (Schackelford and Weekes-Shakelford 2012). Biocultural evolution then has favoured an evolved psychology prescribing not just a genetic disposition to form cooperative alliances/affiliations with other in-group members (as discussed above), but also a disposition to deploy these alliances effectively for defending against hostile neighbouring out-groups. But in turn, cultural selection informing social learning that fosters a penchant for warfare also modulates this natural selection in favouring genotypes that promote even stronger cooperative in-group alliances—because the latter promotes still greater success in warfare. In other words, culture shapes the gene pool, while the gene pool also shapes the culture (Box 6.4). Thus, as Turchin (2016) describes it '… 10,000 years of war made humans the greatest cooperators on earth'—and evidently also played a crucial role in minimizing inequality within societies (Scheidel 2017).

Propensity for lethal violence against members of the same species is measurably greater within the great ape lineage compared with other mammals (Gómez et al. 2016). Our closest genetic relatives, chimpanzees, in particular are known to coordinate local group-based missions against neighbouring groups, including with murder. But there is nothing in other species that compares even remotely with the scale of brutality that continues even today in the unspeakable atrocities of human wars, massacres, torture, ethnic cleansings, and genocides. Smith (2007) puts it effectively into context:

> Like all living things, Homo sapiens possesses an ancient heritage; over the course of many millions of years, the forces of evolution have honed and sculpted our minds and bodies, and this patrimony has an enormous impact on how we live our lives today. The genetic programming bequeathed to us by our ancestors has many constructive, life-affirming aspects. It incites us to seek attractive mates, to savour the flavour of nourishing food, to nurture our children, to understand and control the world around us, and even to compose exquisite music and create wonderful works of art. But our evolutionary legacy also has a much more disturbing face: it moves us to kill our fellow human beings. Violence has followed our species every step of the way in its long journey through time. From the scalped bodies of ancient warriors to the suicide bombers in today's newspaper headlines, history is drenched in human blood.

War is an organized conflict between 'us' versus 'them', usually fueled by deeply ingrained human biases, especially nationalism, racism, xenophobia, and ethnocentrism. For our ancestors, victory of course favoured success for the social groups that they belonged to, and hence also their gene transmission success. It secured more resources and wealth to support the survival of 'us'; raiding 'them' was always preferable to starvation. And when raiding proved to work, it could then be seen as an effective tool for feeding the greed and strengthening the power of corrupt and ruthless leaders. In cases where 'them' espoused a different cultural (e.g. religious) worldview, they sometimes posed a threat to the foundation of what defined a sense of tribal identity and personal meaning for 'us'—including the basis for mortality and self-impermanence anxiety buffers (see Chaps. 9 and 10). As Becker (1971) put it: 'One culture is always a potential menace to another because it is a living example that life can go on heroically within a value framework totally alien to one's own'. Attacking and defeating 'them' thus instilled confidence in local beliefs and bolstered self-esteem by quelling any worries that 'their' cultural worldviews might be superior to 'ours'. 'Men never do evil so completely and cheerfully as when they do it from religious conviction' (Blaise Pascal 1669, Pensées). The massive human slaughters of the Christian crusades, spanning the eleventh to the thirteenth centuries, are a prime example. Solomon et al. (2004) elaborate:

> ... when we do encounter people with different beliefs, this poses a challenge to our death-denying belief systems, which is why people are generally quite uncomfortable around, and hostile towards, those who are different. Additionally, because no symbolic cultural construction can actually overcome the physical reality of death, residual anxiety is unconsciously projected onto other group(s) of individuals as scapegoats: designated as all-encompassing repositories of evil, the eradication of which would make earth as it is in heaven. We then typically respond to people with different beliefs or scapegoats by berating them, trying to convert them to our system of beliefs, and/or just killing them and in so doing assert that 'my God is stronger than your God and we'll kick your ass to prove it.'

For males in particular, going to war also secured access to more females for reproductive subjugation, including rape (Yoshimi 2001) (which would have been of particular interest to young mate-less, sex-starved males in polygynous societies)—especially if combined with the practice of killing or enslaving males in the invaded population (e.g. Marshall 2018). War (if one survived it) could also provide males with a heroic reputation as a 'brave warrior' within the home clan, thus earning status and greater attractiveness to mates, and also appealing to the intrinsic attraction to legacy (as a self-impermanence anxiety buffer; see Chap. 10).

It is easy to consider therefore how these advantages could have favoured an instinctual behavioural penchant for war, especially in males. This does not imply, of course, that men went to war because they specifically wanted to spread their genes. All that mattered is that going to war did indeed serve to spread genes, and all that was needed to accomplish this was an effective behavioural motive for going to war. A motive that has no relationship to any cognitive desire for spreading genes can work just as well—in terms of promoting fitness—as one that does. Even when faced with certain death in war, men have been known to find motivation from being promised access to multiple females in the afterlife. In striking contrast, females have, of course, never been attracted to a suicide bombing assignment by being promised sexual liaisons with multiple males (or any for that matter) in the afterlife.

War then is almost exclusively a 'guy thing', probably because violence itself is. Barash (2008) explains:

> *In the animal world, human no less than nonhuman, competition is often intense. Males typically threaten, bluff, and if necessary fight each other in their efforts to obtain access to females. Among vertebrates in particular, males tend to be relatively large, conspicuous in colour and behaviour, endowed with intimidating weapons (tusks, fangs, claws, antlers, etc) and a willingness to employ them, largely because such traits were rewarded over evolutionary time, with enhanced reproductive success. ... Compared to females, as a result, males tend to be large, fierce, nasty, sneaky, and highly adapted to outmuscle, out- shine, and occasionally even outwit their rivals. Male-male competition is especially fierce in polygynous, harem-keeping species such as elk, moose, elephant seals, or gorillas. ... Homo sapiens are also typically mammalian in their predisposition to polygyny ... polyg- yny was the preferred domestic system for more than 80 percent of human societies.*

Young men, in their late teens and early twenties, have a particular penchant for strutting their stuff with risk-taking and violent displays (the 'young male syn- drome'), because this is when they are entering the 'mating arena'. In civilian life, this predisposes them to a particularly high likelihood of homicide victimization (Wilson and Daly 1985), but in war it tends to make them better-equipped than women for killing. [It also explains why other members of the so-called killing establishment (Barash 2008)—e.g. executioners, hunters, slaughterhouse work- ers—are overwhelmingly male.] But living with oneself as a killer in war requires more. Smith (2007) explains:

> *The anti-parasite module plays an extremely important role in the dehumanizing of enemies of war. ... Dehumanizing the enemy in warfare draws on ancient biological dispositions to overcome the problem posed by the taboo on killing members of our own species. To do this, particular mental modules are activated which cause the soldier to perceive his enemies as human in form but lacking a truly human essence. ... We can regard this as a form of self- deception. In a sense, the soldier must lie to himself about what he is doing. He is not spill- ing the blood of others, he is killing an evil beast, or shooting turkeys, or ridding the world of a terrible disease. ... It follows that when the anti-parasite module is activated and turned against fellow human beings, the stage is set for genocide.*

Other animals spend their whole lives just trying to get fed, stay alive, and get laid. That's about it. The first bi-pedal apes of about 5 million years ago undoubt- edly had similarly defined motivations. And of course, humans today have inherited these same primitive instincts for survival and sex/mating, manifesting in many characteristic human behaviours and cultures (Buss 2009), several of which are briefly reviewed in this and the previous chapter. But the needs of modern humans also involve much more. As French author and philosopher, Albert Camus (1913–1960) observed: 'We humans are creatures who spend our whole lives trying to convince ourselves that our existence is not absurd' (quoted from Tavris and Aronson 2007). Only humans evolved an intrinsic sentiment for possessing an 'inner self', a mental life, and a supposition that it can transcend the mortality of material life—but also an impulsive worry about whether this is so, and thus suscep- tibility to acute anxiety because of it (Chap. 4). This deeply ingrained 'self-imper- manence anxiety' accounts for two additional human motivational domains explored in the next two chapters.

References

Aarssen LW (2009) Not my brother's keeper: a thought experiment for Hamilton's rule. Biosci Hypotheses 2:198–204

Aarssen LW (2013) Will empathy save us? Biol Theory 7:211–217

Al-Shamahi E (2021) The handshake: a gripping history. Profile Books, London

Alexander RD (1985) A biological interpretation of moral systems. Zygon 20:3–20

Barash DR (2008) Natural selections: selfish altruists, honest liars, and other realities of evolution. Bellevue Literary Press, New York

Batthyany A, Russo-Netzer P (eds) (2014) Meaning in positive and existential psychology. Springer, New York

Becker E (1971) The birth and death of meaning: an interdisciplinary perspective on the problem of man, 2nd edn. The Free Press, New York

Bergson H (2020) Laughter is vital: for philosopher Henri Bergson, laughter solves a serious human conundrum: how to keep our minds and social lives elastic. Aeon, 7 July 2020, https://aeon.co/essays/for-henri-bergson-laughter-is-what-keeps-us-elastic-and-free

Bernard LC, Mills M, Swenson L, Walsh RP (2005) An evolutionary theory of human motivation. Genet Soc Gen Psychol Monogr 131:129–184

Bloom P (2013) Just babies: the origins of good and evil. Random House, New York

Boehm C (2012) Moral origins: the evolution of virtue, altruism, and shame. Basic Books, New York

Bowles S, Gintis H (2011) A cooperative species: human reciprocity and its evolution. Princeton University Press, Princeton

Brown DE (2004) Human universals, human nature, and human culture. Daedalus 133:47–54

Buss D (2009) The great struggles of life: Darwin and the emergence of evolutionary psychology. Am Psychol 64:140–148

Cartwright J (2008) Evolution and human behaviour: Darwinian perspectives on human nature, 2nd edn. MIT Press, New York

Cave S (2014) Not nothing: the death of a fly is utterly insignificant – or it's a catastrophe. How much should we worry about what we squash? Aeon Mag. http://aeon.co/magazine/philosophy/how-much-should-we-worry-about-death/

Cosmides L, Tooby J (2013) Evolutionary psychology: new perspectives on cognition and motivation. Annu Rev Psychol 64:201–209

Curtis V (2013) Don't look, don't' touch, don't eat: the science behind revulsion. The University of Chicago Press, Chicago

Danvers A (2021) If smiles are so easy to fake, why do we trust them? Psyche, 23 Feb 2021, https://psyche.co/ideas/if-smiles-are-so-easy-to-fake-why-do-we-trust-them

Darwin C (1859) On the origin of species. Facsimile of the first edition. Harvard University Press, Cambridge

Darwin C (1872) The expression of emotions in man and animals. Philosophical Library, New York

Davis H, McLeod SL (2003) Why humans value sensational news. Evol Hum Behav 24:208–216

Deci EL, Ryan RM (2000) The "what" and "why" of goal pursuits: human needs and the self-determination of behavior. Psychol Inq 11:227–268

Douglas K (2017) Effortless thinking: why stereotyping is an evolutionary trap. New Sci, 13 Dec 2017, https://www.newscientist.com/article/mg23631560-600-effortless-thinking-why-stereotyping-is-an-evolutionary-trap/

Dweck CS, Grant H (2008) Self-theories, goals, and, meaning. In: Shah JY, Gardner WL (eds) Handbook of motivational science. Guilford Press, New York, pp 405–416

Gómez J, Verdú M, González-Megías A, Méndez M (2016) The phylogenetic roots of human lethal violence. Nature 538:233–237

Graziano M (2014) The first smile. Aeon, 13 Aug 2014, https://aeon.co/essays/the-original-meaning-of-laughter-smiles-and-tears

Greenberg J, Solomon S, Arndt J (2008) A basic but uniquely human motivation: terror management. In: Shah JY, Gardner WL (eds) Handbook of motivational science. Guilford Press, New York, pp 114–134

Hare B, Woods V (2020) Survival of the friendliest: understanding our origins and rediscovering our common humanity. Random House, New York

Heine SJ, Proulx T, Vohs KD (2006) The meaning maintenance model: on the coherence of social motivations. Personal Soc Psychol Rev 10:88–110

Hendriksen E (2018) Why everyone is insecure (and why that's okay). Sci Am, April 12, 2018, https://blogs.scientificamerican.com/observations/why-everyone-is-insecure-and-why-thats-okay/

Henrich N, Henrich J (2007) Why humans cooperate: a cultural and evolutionary explanation. Oxford University Press, New York

Huang JY, Bargh JA (2014) The selfish goal: autonomously operating motivational structures as the proximate cause of human judgment and behavior. Behav Brain Sci 37:121–175

Jackson RE, Cormack LK (2007) Evolved navigation theory and the descent illusion. Percept Psychophys 69:353–362

Jordan JJ, Hoffman M, Bloom P, Rand DG (2016) Third-party punishment as a costly signal of trustworthiness. Nature 530:473–476

Joyce R (2006) The evolution of morality. MIT Press, Cambridge

Kelly DJ, Quinn PC, Slater AM, Lee K, Gibson A, Smith M, Ge L, Pascalis O (2005) Three-month-olds, but not newborns, prefer own-race faces. Dev Sci 8:F31–F36

Keltner D, Gross JJ (1999) Functional accounts of emotions. Cognit Emot 13:467–480

Kenrick DT, Griskevicius V (2013) The rational animal: how evolution made us smarter than we think. Basic Books, New York

Kenrick DT, Griskevicius V, Neuberg SL, Schaller M (2010) Renovating the pyramid of needs: contemporary extensions built upon ancient foundations. Perspect Psychol Sci 5:292–314

Kesebir S, Graham J, Oishi S (2010) A theory of human needs should be human-centered, not animal-centered: commentary on Kenrick et al. (2010). Perspect Psychol Sci 5:315–319

Knight C (2019) Did laughter make the mind? A psychological relief valve and a guard against despotism, laughter is a uniquely human – and collective – activity. Aeon, 11 Feb 2019, https://aeon.co/essays/does-laughter-hold-the-key-to-human-consciousness

Kteily N, Bruneau E, Waytz A, Cotterill S (2015) The ascent of man: a theoretical and empirical case for blatant dehumanization. J Pers Soc Psychol 109:901–931

Lawlor R (2012) The ethical treatment of animals: the moral significance of Darwin's theory. In: Brinkworth M, Weinert F (eds) Evolution 2.0. Implications of Darwinism in philosophy and the social and natural sciences. Springer, Berlin

Levy N (2019) Is virtue signalling a perversion of morality? Aeon, 29 Sept 2019, https://aeon.co/ideas/is-virtue-signalling-a-perversion-of-morality

Li YJ, Kenrick DT, Griskevicius V, Neuberg SL (2012) Economic decision biases and fundamental motivations: how mating and self-protection alter loss aversion. J Pers Soc Psychol 102:550–561

LoBue V, DeLoache JS (2010) Superior detection of threat-relevant stimuli in infancy. Dev Sci 13:221–228

Marshall M (2018) Every man in Spain was wiped out 4500 years ago by hostile invaders. New Sci, 28 Sept 2018, https://www.newscientist.com/article/2180923-every-man-in-spain-was-wiped-out-4500-years-ago-by-hostile-invaders/

Maslow AH (1943) A theory of human motivation. Psychol Rev 50:370–396

McAndrew FT (2008) The science of gossip: why we can't stop ourselves. Sci Am Mind, October, pp 26–33

Navarrete CD, Fessler DMT, Eng SJ (2007) Elevated ethnocentrism in the first trimester of pregnancy. Evol Hum Behav 28:60–65

Nettle D (2009) Ecological influences on human behavioural diversity: a review of recent findings. Trends Ecol Evol 24:618–624

Orians GH (2014) Snakes, sunrises, and Shakespeare: how evolution shapes our loves and fears. The University of Chicago Press, Chicago

Peterson C, Park N (2010) What happened to self-actualization? Commentary on Kenrick et al. (2010). Perspect Psychol Sci 5:320–322

Ridley M (1998) The origins of virtue: human instincts and the evolution of cooperation. Penguin Books, New York

Sarner M (2017) Feeling lonely? You're not on your own. New Sci, 19 July 2017, https://www. newscientist.com/article/mg23531351-900-feeling-lonely-youre-not-on-your-own/

Schackelford TK, Weekes-Shakelford VA (2012) The Oxford handbook of evolutionary perspective on violence, homicide, and war. Oxford University Press, Oxford

Scheidel W (2017) The great leveler: violence and the history of inequality from the stone age to the twenty-first century. Princeton University Press, Princeton

Schnell T (2012) An existential turn in psychology. Meaning in life operationalized. Habilitation Treatise. University of Innsbruck, Innsbruck

Shah JY, Gardner WL (eds) (2008) Handbook of motivational science. Guilford Press, New York

Shaver PR, Mikulincer M (eds) (2012) Meaning, mortality, and choice: the social psychology of existential concerns. American Psychological Association, Washington, DC

Sherman PW, Billing J (1999) Darwinian gastronomy: why we use spices. Bioscience 49:453–463

Slingerland E (2021) Drunk: how we sipped, danced, and stumbled our way to civilization. Little, Brown Spark, New York

Smith DL (2007) The most dangerous animal: human nature and the origins of war. St. Martin's Press, New York

Solomon S, Greenberg JL, Pyszczynski TA (2004) Lethal consumption: death-denying materialism. In: Kasser T, Kanner AD (eds) Psychology and consumer culture: the struggle for a good life in a materialistic world. American Psychological Association, pp 127–146

Sterelny K (2012) The evolved apprentice: how evolution made humans unique. MIT Press, Cambridge

Tarr B (2017) Social bonding through dance and 'musiking'. In: Enfield NL, Kockelman P (eds) Distributed agency. Oxford University Press, London, pp 151–158

Tavris C, Aronson E (2007) Mistakes were made (but not by me): why we justify foolish beliefs, bad decisions, and hurtful acts. Harcourt, Orlando

Tomasello M (2009) Why we cooperate. MIT Press, Cambridge

Tomasello M (2018) The origins of human morality: how we learned to put our fate in one another's hands. Sci Am 319:70–75

Turchin P (2016) Ultra society: how 10,000 years of war made humans the greatest cooperators on earth. Beresta Books, Chaplin

van Bruggen V, Vos J, Westerhof G, Bohlmeijer E, Glas G (2015) Systematic review of existential anxiety instruments. J Humanist Psychol 55:173–201

Van Prooijen J-W (2019) Suspicion makes us human. Aeon, 4 Nov 2019, https://aeon.co/essays/ how-conspiracy-theories-evolved-from-our-drive-for-survival

Vess M, Routeldge C, Landau MJ, Arndt J (2009) The dynamics of death and meaning: the effects of death-relevant cognitions and personal need for structure on perceptions of meaning in life. J Pers Soc Psychol 97:728–744

Wilson EO (1978) On human nature. Harvard University Press, Cambridge

Wilson M, Daly M (1985) Competitiveness, risk taking, and violence: the young male syndrome. Ethol Sociobiol 6:59–73

Wrangham R (2019) The goodness paradox: the strange relationship between virtue and violence in human evolution. Vintage Books, New York

Yoshimi Y (2001) Comfort women: sexual slavery in the Japanese military during World War II. Columbia University Press, New York

Chapter 9
Escape from Self

Man staggers through life yapped at by his reason, pulled and shoved by his appetites, whispered to by fears, beckoned by hopes. Small wonder that what he craves most is self-forgetting. (Eric Hoffer 1955)

Pamela Coleman Smith (1909) *The Fool*/Wikipedia/Public Domain

Building on inspiration from the Kenrick et al. (2010) 'pyramid of needs' renovation (Box 8.1), I offer here another version for remodeling a pyramid of human 'drives' (Box 9.1)—also conceptually framed by Darwinian evolution. To reinforce the central importance of the latter, the exalted pyramid cap represents not a motivation per se, but the overarching functional (adaptive) consequence connected to all of the underlying needs/drives: successful transmission of gene copies into future generations. The lowest and highest categories of motivations in Box 9.1 have essentially the same elements as corresponding levels in the Kenrick et al. version. In the latter, those levels associated with the core of 'somatic effort'—Immediate Physiological Needs, Self-Protection, and Affiliation—are subsumed here under Survival Drive (Box 9.1), considered in Chap. 8. Similarly, the higher order 'reproductive effort' needs—Mate Acquisition, Mate Retention, and Parenting needs from Box 8.1—are distilled here within Sexual/Familial Drives, which also includes kin-helping (Box 9.1), considered in Chap. 7 (and 8).

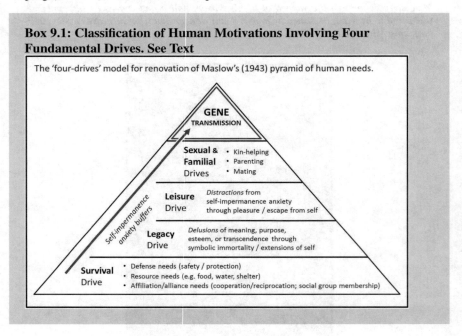

Box 9.1: Classification of Human Motivations Involving Four Fundamental Drives. See Text

The 'four-drives' model for renovation of Maslow's (1943) pyramid of human needs.

GENE TRANSMISSION

Sexual & Familial Drives
• Kin-helping
• Parenting
• Mating

Leisure Drive — *Distractions* from self-impermanence anxiety through pleasure / escape from self

Legacy Drive — *Delusions* of meaning, purpose, esteem, or transcendence through symbolic immortality / extensions of self

Self-impermanence anxiety buffers

Survival Drive
• Defense needs (safety / protection)
• Resource needs (e.g. food, water, shelter)
• Affiliation/alliance needs (cooperation/reciprocation; social group membership)

The most significant renovation proposed here involves greater emphasis on a 'narrative of the self', involving motivational elements that include and expand on notions of personal identity and self-actualization—not as a priority (in the Maslow sense; Chap. 8) only after all other core needs are satisfied, but rather as a more integrative construct at the mid-level. The notion of humans having a pyramid of 'needs' or 'drives' is nothing short of a model of self-interest emerging from our evolution of self-salience (Chap. 4). Some of this of course is about 'pure' selfishness, but not all of it or even much of it. As we saw in Chap. 8, evolutionary theory interprets degrees of self-interest that are often blended subtly and strategically with adaptive interest in, and benevolence for, other in-group members. This relationship was elegantly recognized, prior to Darwin, over two centuries ago in a book by English essayist William Hazlitt (1805)—which until recently was largely overlooked by social theorists (Martin and Barresi 1995; Barbalet 2009; Savarese 2013). Hazlitt (1805) writes:

> *It is by comparing the knowledge that I have of my own impressions, ideas, feelings, powers, etc. with my knowledge of the same or similar impressions, ideas, etc. in others, and with the still more imperfect conception that I form of what passes in their minds when this is supposed to be essentially different from what passes in my own, that I acquire the general notion of self. If I had no idea of what passes in the minds of others, or if my ideas of their feelings and perceptions were perfect representations, i.e., mere conscious repetitions of them, all proper personal distinction would be lost either in pure self-love, or in perfect universal sympathy.*

Following on the premise of needs/drives as 'self-interest', 'Legacy Drive' is identified here as a distinct intermediary domain of motivation (Box 9.1)—and within which, the 'Status/Esteem' needs of Kenrick et al. (2010) are subsumed. 'Legacy Drive' (considered in Chap. 10), together with a second intermediary domain—'Leisure Drive' (Box 9.1), considered below—represent adaptive cognitive systems rooted in uniquely human, innate predilections for existential meaning, purpose, and larger-than-self identity (Aarssen 2010; Klinger 2012; MacKenzie and Baumeister 2014). Importantly, it is argued here, failure to meet these domain-level psychological needs—as with failure to meet survival needs—imposed severe ancestral limitations on deployment of the higher order (Sexual/Familial) Drives linked most proximally to gene transmission success.

Leisure Drive

Leisure is an integral component of the human pursuit of happiness and wellbeing (Newman et al. 2014), with roots in Epicureanism, the 'pleasure principle' of Freudian psychology, and the 'nirvana' of Buddhism. It has long been an institution unto itself, with whole journals dedicated to its study in sociology and social psychology (e.g. Tinsley and Tinsley 1986; Crawford et al. 1991; Iwasaki 2008; Kleiber et al. 2011; Brajsa-Zganec et al. 2011; Barnett 2013; Freire 2013; Gander et al. 2013). Other animals might be considered to engage in leisure, but if they do, the scale, complexity, and time/energy/material allocation commitments involved all pale in comparison to humans. As Oscar Wilde (1891) wrote, 'Pleasure is the only thing worth having a theory about'.

Attraction to leisure, 'Leisure Drive', is regarded here as a deeply ingrained adaptive cognitive domain, defining a fundamental human need to dispel negative mood/feelings, and hence to find release from suffering and freedom from pain, deployed particularly as a distraction, and hence, reprieve from the chronic and uniquely human worry that one's mental life (the 'inner self') is—like one's material life—also impermanent (Aarssen 2007, 2010; see also Clack 2014) (Chap. 4). Leisure Drive is satisfied by hard-wired pleasure perceptions associated with various forms of sensory/sensual gratification/euphoria—e.g. involving relaxation, cognitive stimuli, intellectual enlightenment, social affiliations/entertainments, physical fitness and exercise, recreational athletics and team sports, and various positive emotions connected with intimacy, love, empathy, family relations, companionship, belongingness, etc. Initially, pleasure-based motivations to seek and engage with these experiences—modulated by 'cocktails' of hormones and neurotransmitters—would have evolved because of their particular rudimental or proximate effects on evolutionary fitness (Martin 2008; Bloom 2010): i.e. in promoting survival (Chap. 8; e.g. resting; eating nutritious foods; cultivating friendships and social alliances; learning and understanding through curiosity and imagination; using pretending, play, games, and sports to sharpen social intelligence, practice social skills, and to rehearse for war)—and in promoting reproduction (Chap. 7; e.g. sexual gratification, courtship, romantic love, parental love, affection for kin, and hence welfare for one's collateral lineage).

Importantly however, as posited here, natural selection was not finished. Attraction to the above activities and pursuits was honed further by evolution, *purely for the sake of their sheer intrinsic enjoyment.* In other words, biocultural evolution (Chap. 6) gave us Leisure Drive by 'hacking' into these primal, innately endowed enjoyment modules anticipated in the daily routines of our ancestors, and deploying them within cultures of *intentionally planned and prioritized free-time indulgence*— i.e. directly functional as a domain-general need in its own right. As William James (1902) asserted: 'How to gain, how to keep, how to recover happiness is in fact for most men at all times the secret motive of all they do, and of all they are willing to endure'. As ventures of leisure, these enjoyment modules would have served a vital role in propelling ancestral gene copies into future generations: *by providing buffers from the inevitable agonies of human existence, including especially the most elemental, and most debilitating of all: minds troubled by the chronic worry of self-impermanence* (Chap. 10)—thus enabling ancestors and their descendants (us) to reap the profound fitness benefits of mental time travel and theory of mind (Chap. 4).

Mortality salience and 'passage of time' salience have been shown previously to activate thoughts of enjoying one's life (Mogilner 2010). 'Forgetting of self' and 'decreased awareness of time' are thus regarded as key cognitive attributes of leisure experiences (Tinsley and Tinsley 1986), to which we can thus add—in addition to those above—mind-altering escapism in cultures like stories, films/videos, theatre, television, comedy, carnivals, exhibitions, festivals, travel, amusement parks, parties, raves, thrill-seeking, shopping, toys, hobbies, vacations, aesthetics, music, art, nostalgia, yoga, meditation, prayer, secular spiritualism/mysticism, mindfulness, sleep, daydreams, mood medications, snacking, recreational sex, intoxication, and psychedelics. Such palliatives are evident in some of the oldest written records: 'Give strong drink to him who is perishing, and wine to those who are bitter of heart. Let him drink and forget his poverty, and remember his misery no more' (Proverbs 31:6–7; New King James Version). And from eleventh-century Persian poet and mathematician, Omar Khayyam: 'I drink not from mere joy in wine nor to scoff at faith—no, only to forget myself for a moment—that only do I want of intoxication, that alone' (quoted from Becker 1973). As Louis Lewin (1998) put it: 'A man must sometimes take a rest from his memory'.

Many of our treasured hobbies are obvious representations of enjoyment modules that rewarded the reproductive success of our distant ancestors—e.g. associated with obtaining food (e.g. hunting, fishing, gardening, gourmet cooking), obtaining shelter (e.g. camping, wood-working, quilt-making, knitting), and exploration (e.g. hiking, canoeing, sailing, tourism). Opportunities for, and discoveries/technologies supporting, the above indulgences have grown exponentially ever since the dawn of Neolithic agriculture—through biocultural evolution, driven and shaped to a large extent I suggest because of Leisure Drive. We are descendants of the brave and ambitious, and also the devout, but we are also descendants of the inventors and consumers of hobbies, games, meditation, art, and beer. As Dutton (2009) put it:

> The evolution of Homo sapiens in the past million years is not just a history of how we came to have acute colour vision, a taste for sweets, and an upright gait. It is also a story of how we became a species obsessed with creating artistic experiences with which to amuse, shock, titillate, and enrapture ourselves, from children's games to the quartets of Beethoven, from firelit caves to the continuous worldwide glow of television screens.

Biocultural evolution has thus generated a large 'leisure menu' for us to choose from, and the remedy of choice will of course vary according to individual tastes, and perceptions of one's personal strengths and potential. By distracting the mind with these many options for triggering pleasure (contentment, joy, happiness, bliss, euphoria, ecstasy, rapture, nirvana)—with its attendant displays of smiles and laughter [which can themselves evoke feedback effects in reinforcing happiness (Wenner 2009)]—Leisure Drive then enables a forgetting of mortality, and hence a denial of death (Becker 1973). It delivers an escape from time and from the terror of history (Ruiz 2011) by palliating the many hardships, sufferings, and absurdities that (for most of ancestral as well as contemporary humanity) were/are routine consequences of just being alive. These distractions/diversions (occurring with sufficient magnitude, duration, and frequency), it can be argued, thus promoted our ancestors' genetic fitness by preventing self-impermanence salience and other anxieties from compromising the successful deployment of Sexual/Familial Drives (Box 9.1).

After all, for some people, 'escape from self' also has a darker side, as motivation for suicide (Baumeister 1990; Bering 2018), which of course is not in the best interests of transmission for 'selfish' genes, especially if enacted prior to reproductive maturity. 'Discovery of self' (Chap. 4) had enormous fitness benefits for our ancestors—but the trade off, the 'burden of discovery' (self-impermanence anxiety), while manageable for most—has probably always been just too much to bear for a minority. As Humphrey (2018) put it, 'Nothing hurts less than being dead'.

Recent studies have shown that when the mind is allowed to wander disengaged, essentially unoccupied by fulfilling motivations and unprotected by 'present-moment' mindfulness, it is usually an anxious, unhappy mind (Killingsworth and Gilbert 2010; Wilson et al. 2014). Leisure Drive however functions (possibly subconsciously much of the time) to help ensure that such anxieties are kept largely in check by pleasurable ventures for escape to 'outside-of-self', e.g. through 'flow states' (Csikszentmihalyi 1975)—or by remedies that provide versions of mental oblivion less lethal than suicide, e.g. 'altered states' or 'out-of-body-experiences' (Gandy et al. 2020), involving induced sleep, sedation, anaesthesia, hypnosis, sudden amnesia, inebriation, and narcosis (Aarssen 2018; Hales 2021). With deployment of these options, it is perhaps not difficult to understand how the agonies of existence might persuade some that it would be 'better never to have been' (Benatar 2006, 2017). From Blaise Pascal (1670; *Pensées*):

> *The only thing which consoles us for our miseries is diversion, and yet this is the greatest of our miseries. For it is this which principally hinders us from reflecting upon ourselves, and which makes us insensibly ruin ourselves. Without this we should be in a state of weariness, and this weariness would spur us to seek a more solid means of escaping from it. But diversion amuses us, and leads us unconsciously to death.*

And from William Hazlitt (1815):

> *We are not going to enter into the question, whether life is, on the whole, to be regarded as a blessing, though we are by no means inclined to adopt the opinion of the sage who thought "that the best thing that could have happened to a man was never to have been born, and the next best to have died the moment after he came into existence.*

Depression and other forms of mental illness thus remain relatively common, but some writers have suggested that some of these disorders have components that were probably adaptive for at least some of our ancestors, e.g. in terms of greater capacity for creativity, and capacity to assess and solve complex problems (Andrews and Thompson 2009; Ghaemi 2011; Ravilious 2011; Nesse 2019). From Friedman (2014):

> *Consider that humans evolved over millions of years as nomadic hunter-gatherers. It was not until we invented agriculture, about 10,000 years ago, that we settled down and started living more sedentary—and boring—lives. As hunters, we had to adapt to an ever-changing environment where the dangers were as unpredictable as our next meal. In such a context, having a rapidly shifting but intense attention span and a taste for novelty would have proved highly advantageous in locating and securing rewards—like a mate and a nice chunk of mastodon. In short, having the profile of what we now call A.D.H.D. would have made you a Paleolithic success story.*

In some cases therefore, diagnosis as a 'disorder' may be a misnomer. Anxiety and depression, for example, may commonly be just ordinary consequences of having a particularly acute grasp of reality—a heritable cognitive tool likely to have been essential for survival in Pleistocene Africa (Syme and Hagen 2020). But in the modern human mind, an acute grasp of reality can also lead to perfectly 'normal' existential angst. Choron (1964) cites an interesting account in this regard from the Austrian psychiatrist Victor Frankl (1959): 'Man's will-to-meaning represents the most human phenomenon possible, and its frustration does not signify something pathological, at least not in itself. A person is not necessarily sick if he thinks that his existence is meaningless. A doctor who would confuse a human phenomenon with a pathological symptom ... would be prone to drown the spiritual distress—the existential frustration—of his patient in gallons of tranquilizers'. Nevertheless, it seems reasonable to speculate that, in some cases, individuals might be inadequately protected from anxieties evoked by a lost sense of meaning—i.e. insufficiently equipped for deployment of distractions (through Leisure Drive) or delusions (through Legacy Drive; Chap. 10). Such cases, however, have recently shown a troubling increase in frequency. We return to this topic in the final chapter.

Conspectus

According to the great Russian author, Leo Tolstoy (1872): 'For man to be able to live he must either not see the infinite, or have such an explanation of the meaning of life as will connect the finite with the infinite'. Tolstoy evidently implies here that a life can have 'meaning' if one is able to embrace an 'explanation' for it that is grounded firmly in an unwavering sense of 'self-hood'—a mental life—that can exist apart from the material body and so, unlike the latter, can endure/transcend beyond corporeal existence (Chap. 4), i.e. 'connecting the finite with the infinite'. Ostensibly therefore, this evokes a necessary condition for meaning: confidence (albeit grounded in a steadfast delusion) that one's existence is not absurd—which, as Albert Camus famously proclaimed, we spend our whole lives trying to be convinced of (Aarssen 2018). 'We humans have purpose on the brain' (Dawkins 1995) (see Chap. 10).

But Tolstoy also implies that there is an alternative sense in which a life can have purpose or 'meaning': by 'not seeing the infinite'. As argued in this chapter, this is enabled by an evolved psychology providing maintenance of an untroubled mind through intentional deployment of triggers for enjoyment [including planning for future deployment (e.g. Iigaya et al. 2020)] as effective distractions from eventual mortality/self-impermanence anxiety, and thus from the nagging worry that there possibly/probably isn't any 'meaning' at all—that one's existence might, in fact, be absurd. Preoccupation, activated by Leisure Drive, effectively diverts the sentient mind from its impulsive tendency to arrive at (and to relentlessly revisit) an unsettling deduction: *that notions of self-transcendence or of a grand cosmic purpose for one's life are/may be nothing but imaginary mental constructs.*

Leisure Drive then is a built-in cognitive tool for mental catharsis, a therapy for evoking 'positive thinking'—a kind of placebo elixir for calming the troubled mind. It doesn't matter that the calming is achieved by distraction; all that matters is that it commonly works—by liberating the mind from its own conflicted sense of 'self'. And it doesn't matter that the diversion, for some, may require a busy mind; all that matters is that the mind is not troubled. People with a strong Leisure Drive are, by definition, optimistic, and recent research shows that optimism is 'in our genes' (Fox et al. 2009), hence, for example, accounting for the success of 'feel good' books like the best-seller, *The Book of Awesome* (Pasricha 2010).

In other words, the 'purpose' here is in the pleasure—or more specifically, the serotonin/oxytocin/endorphin/dopamine/endocannabinoid rushes that facilitate it. It doesn't need to work all of the time—just often enough. For many then, a life may be meaningful by simply filling it periodically with things that feel good, thus taking the mind to a place 'outside-of-self', by 'arresting' it temporarily but firmly in the bliss of the immediate present, hence effectively shielding it from the dread of the hereafter—Tolstoy's 'not seeing the infinite'—wherein the self essentially ceases to be. The rapture of awe and the sublime of wonder—wherein one's 'breath is taken away' by reading something powerful and utterly profound, or by viewing something vast in scale, or something magnificently beautiful or serenely peaceful, or by the epistemic ecstasy of a novel idea or landmark discovery, emerging from raw curiosity—are particularly cogent illustrations of this glorious distraction (Marchant 2017; Chirico and Yaden 2018; De Cruz 2020). As Rachel Carson (1956) wrote, 'Those who dwell, as scientists or laymen, among the beauties and mysteries of the Earth are never alone or weary of life. Whatever the vexations or concerns of their personal lives, their thoughts can find paths that lead to inner contentment and to renewed excitement in living'. A rumination from poet Wendell Berry (2012)— *The Peace of Wild Things*—is particularly expressive:

When despair for the world grows in me
and I wake in the night at the least sound
in fear of what my life and my children's lives may be,
I go and lie down where the wood drake
rests in his beauty on the water, and the great heron feeds.
I come into the peace of wild things
who do not tax their lives with forethought
of grief. I come into the presence of still water.
And I feel above me the day-blind stars
waiting with their light. For a time
I rest in the grace of the world, and am free.

Effective domains of escape from self, however, need not be particularly exhilarating. In many cases, they may be pleasant enough (e.g. restfully reclining on the sofa while snacking and mindlessly flicking through TV channels or scrolling through Facebook pages on a smartphone) simply because they bring a welcome solace: diversion from the curse of self-impermanence anxiety. The role of leisure here then is, most essentially, palliative. It delivers respite—a 'feel good' vehicle for periodic escape—from the deeply ingrained sentiment of having a distinct 'inner self', separate from 'material life' (Chap. 4), and hence respite from the exasperating human obsession of striving to be convinced that one's existence is not absurd. The act of 'striving' for this is just as true for those whose remedies involve adherence to mindfulness and meditation as it is for those who are drawn more to other (perhaps less edifying) distractions, like toys, stories, games, and good times with friends at the pub—or those who are able to feel good by just getting 'lost in the crowd' (Wisman 2006), or by just 'keeping busy' (Hoffer 1976). Our predecessors who were good at keeping busy, I suggest, probably left more descendants, on average, than those who were good at keeping calm.

Importantly, though, the more edifying and ethical pursuits of 'escape from self' will normally require that others (and oneself) are not harmed and that the environment is not harmed. Longstanding and growing concerns regarding the latter are examined in Chap. 12—and prudence of course is commonly required in order to avert the health risks and societal costs of personal over-indulgence, e.g. in binge eating, or substance abuse. Interestingly though for some, anxiety distraction/buff-ering may be procured through intentional infliction of physical pain (masochism) (Baumeister 1989). The emotions of both pleasure and pain, and also fear (Clasen 2012; Javanbakht and Saab 2017), can apparently trigger a kind of catharsis—a mental 'arrest' in the immediate present, and thus, an escape from 'self', where the mind is temporarily but gloriously shielded from regrets of the past, as well as wor-ries of the future, wherein self-impermanence resides for all eternity.

Nevertheless, for most people today (thanks in part to genetic bequeathal inform-ing empathic/moral/pro-social instincts—see Chap. 8), enjoyment is commonly evoked by helping (not harming) others within one's social group, and from show-ing a spirit of kindness (Dunn et al. 2008; Rudd et al. 2014) and cooperation (Bowles and Gintis 2011). Even when this involves the busyness of hard work, it can be enjoyed not only because it is purposeful—it also delivers therapeutic diversion for a troubled mind. As many wise grandmothers have advised, 'there is always some-one worse off than you. Find them, help them, and you'll feel better'.

All of this, however, begs an underlying question: why is the above deduction—that notions of self-transcendence are probably absurd—so routinely salient and unsettling for the human mind? As William Hazlitt (1822) and other luminaries mused: 'There was a time when we were not: this gives us no concern—why then should it trouble us that a time will come when we shall cease to be?' The answer, argued in the next chapter, is that this whole syndrome of mental anguish generated (ironically) an evolved psychology—a 'hard wired' and uniquely human goal sys-tem—that played a critical role in driving ancestors to produce offspring, or to gen-erate accomplishments that promoted their own survival or reproductive success, or that of their offspring. All animals have a Survival Drive and a Sex Drive. And some other animals (e.g. through what seems like 'play', especially in juveniles) might also enjoy what looks, to us, like a rudimental leisure. But only humans have Legacy Drive.

References

Aarssen LW (2007) Some bold evolutionary predictions for the future of mating in humans. Oikos 116:1768–1778

Aarssen LW (2010) Darwinism and meaning. Biol Theory 5:296–311

Aarssen LW (2018) Meet *Homo absurdus*—the only creature that refuses to be what it is. Ideas Ecol Evol 11:90–95

Andrews PW, Thompson JA (2009) The bright side of being blue: depression as an adaptation for analyzing complex problems. Psychol Rev 116:620–654

Barbalet J (2009) Disinterestedness and self-formation: principles of action in William Hazlitt. Eur J Soc Theory 12:195–211

Barnett L (2013) What people want from their leisure: the contributions of personality facets in differentially predicting desired leisure outcomes. J Leis Res 45:150–191

Baumeister RF (1989) Masochism and the self. Lawrence Erlbaum Associates, Mahwah

Baumeister RF (1990) Suicide as escape from self. Psychol Rev 97:90–113

Becker E (1973) The denial of death. Simon and Schuster, New York

Benatar D (2006) Better never to have been: the harm of coming into existence. Oxford University Press, New York

Benatar D (2017) The human predicament: a candid guide to life's biggest questions. Oxford University Press, New York

Bering J (2018) Suicidal: why we kill ourselves. The University of Chicago Press, Chicago

Berry W (2012) New collected poems. Counter Point, Berkeley. https://aeon.co/videos/the-poet-wendell-berry-reflects-on-the-sublime-peace-of-escaping-into-wilderness

Bloom P (2010) How pleasure works: the new science of why we like what we like. Norton and Company, New York

Bowles S, Gintis H (2011) A cooperative species: human reciprocity and its evolution. Princeton University Press, Princeton

Brajsa-Zganec A, Merkas M, Sverko I (2011) Quality of life and leisure activities: how do leisure activities contribute to subjective well-being. Soc Indic Res 102:81–91

Carson R (1956) Help your child to wonder. Woman's Home Companion, July 1956, pp 25–27, 46–47. https://rachelcarsoncouncil.org/wp-content/uploads/2019/08/whc_rc_sow_web.pdf

Chirico A, Yaden DB (2018) Awe: a self-transcendent and sometimes transformative emotion. In: Lench H (ed) The function of emotions. Springer, Cham

Choron J (1964) Modern man and mortality. The Macmillan Company, New York

Clack BR (2014) Love, drugs, art, religion: the pains and consolations of existence. Ashgate Publishing Ltd., Farnham

Clasen M (2012) Monsters evolve: a biocultural approach to horror stories. Rev Gen Psychol 16:222–229

Crawford DW, Jackson EL, Godbey G (1991) A hierarchical model of leisure constraints. Leis Sci 13:309–320

Csikszentmihalyi M (1975) Beyond boredom and anxiety: experiencing flow in work and play. Jossey-Bass, San Francisco

Dawkins R (1995) River out of Eden: a Darwinian view of life. Basic Books, New York

De Cruz H (2020) The necessity of awe: in awe we hold fast to nature's strangeness and open up to the unknown; no wonder it's central to the scientific imagination. Aeon, 10 July, 2020. https://aeon.co/essays/how-awe-drives-scientists-to-make-a-leap-into-the-unknown

Dunn EW, Aknin LB, Norton MI (2008) Spending money on others promotes happiness. Science 319:1687–1688

Dutton D (2009) The art instinct: beauty, pleasure, and human evolution. Bloomsbury Press, New York

Fox E, Ridgewell A, Ashwin C (2009) Looking on the bright side: biased attention and the human serotonin transporter gene. Proc R Soc B 276:1747–1751

Frankl V (1959) From death-camp to existentialism. Beacon Press, Boston

Freire T (2013) Positive leisure science: from subjective experience to social contexts. Springer, Dordrecht

Friedman RA (2014) A natural fix for ADHD. N Y Times, 31 Oct 2014, http://www.nytimes.com/2014/11/02/opinion/sunday/a-natural-fix-for-adhd.html?action=click&contentCollection=Dining%20%26%20Wine&module=MostEmailed&version=Full&region=Marginalia&src=me&pgtype=article&_r=0

Gander F, Proyer RT, Ruch W, Wyss T (2013) Strength-based positive interventions: further evidence for their potential in enhancing well-being and alleviating depression. J Happiness Stud 14:1241–1259

Gandy S et al. (2020) Altered states can help us face death with serenity and levity. Pysche, 8 Dec 2020, https://psyche.co/ideas/altered-states-can-help-us-face-death-with-serenity-and-levity

Ghaemi N (2011) A first-rate madness: uncovering the links between leadership and mental illness. Penguin Press, New York

Hales S (2021) Sudden amnesia showed me the self is a convenient fiction. Psyche, 11 Jan 2021, https://psyche.co/ideas/sudden-amnesia-showed-me-the-self-is-a-convenient-fiction

Hazlitt W (1805) An essay on the principles of human action: being an argument in favour of the natural disinterestedness of the human mind. J. Johnson, St. Paul's Churchyard, London. https://archive.org/details/anessayonprinci00hazlgoog

Hazlitt W (1815) On the love of life. The Examiner, 15 Jan 1815. From: Keynes G (ed) (2010) Selected essays of William Hazlitt 1778 to 1830. Kessinger Publishing, Whitefish

Hazlitt W (1822) On the fear of death. In: Table talk, essays on men and manners. From: Keynes G (Ed) (2010) Selected essays of William Hazlitt 1778 to 1830. Kessinger Publishing, Whitefish

Hoffer E (1955) The passionate state of mind and other aphorisms. Harper, New York

Hoffer E (1976) The ordeal of change. Buccaneer Books, Cutchogue

Humphrey N (2018) The lure of death: suicide and human evolution. Philos Trans R Soc B 373:20170269

Iigaya K, Hauser TU, Kurth-Nelson Z et al (2020) The value of what's to come: Neural mechanisms coupling prediction error and the utility of anticipation. Sci Adv 6(25):eaba3828. https://doi.org/10.1126/sciadv.aba3828

Iwasaki Y (2008) Pathways to meaning-making through leisure-like pursuits in global contexts. J Leis Res 40:231–249

James W (1902) The varieties of religious experience: a study in human nature. Longmans, Green & Co., London

Javanbakht A, Saab L (2017) The science of fright: why we love to be scared. https://theconversation.com/the-science-of-fright-why-we-love-to-be-scared-85885

Kenrick DT, Griskevicius V, Neuberg SL, Schaller M (2010) Renovating the pyramid of needs: contemporary extensions built upon ancient foundations. Perspect Psychol Sci 5:292–314

Killingsworth MA, Gilbert DT (2010) A wandering mind is an unhappy mind. Science 330:932

Kleiber DA, Walker GJ, Mannell RC (2011) A social psychology of leisure, 2nd edn. Venture Publishing, State College

Klinger E (2012) The search for meaning in evolutionary goal-theory perspective. In: Wong PTP (ed) The human quest for meaning: theories, research, and applications, 2nd edn. Routledge, New York, pp 23–55

Lewin L (1998) Phantastica: a classic survey on the use and abuse of mind-altering plants. Park Street Press, South Paris

MacKenzie MJ, Baumeister RF (2014) Meaning in life: nature. Needs, and myths. In: Batthyany PA, Russo-Netzer P (eds) Meaning in positive and existential psychology. Springer, New York, pp 25–37

Marchant J (2017) Awesome awe: the emotion that gives us superpowers. New Sci, 26 July 2017, https://www.newscientist.com/article/mg23531360-400-awesome-awe-the-emotion-that-gives-us-superpowers/

Martin P (2008) Sex, drugs, and chocolate: the science of pleasure. Fourth Estate, London

Martin R, Barresi J (1995) Hazlitt on the future of the self. J Hist Ideas 56:463–481

Mogilner C (2010) The pursuit of happiness: time, money, and social connection. Psychol Sci 21:1348–1354

Nesse RM (2019) Good reasons for bad feelings: insights from the frontier of evolutionary psychiatry. Dutton, New York

Newman DB, Tay L, Diener E (2014) Leisure and subjective well-being: a model of psychological mechanisms as mediating factors. J Happiness Stud 15:555–578

Pascal B (1670) Pensées and other writings; translation by Honor Levi (1995). Oxford University Press, New York

Pasricha N (2010) The book of awesome. Penguin Group, New York

Ravilious K (2011) Misfit minds gave our ancestors the edge. New Sci 212:34–37

Rudd M, Aakerb J, Nortonc MI (2014) Getting the most out of giving: concretely framing a prosocial goal maximizes happiness. J Exp Soc Psychol 54:11–24

Ruiz TF (2011) The terror of history: on the uncertainties of life in Western civilization. Princeton University Press, Princeton

Savarese J (2013) Reading one's own mind: Hazlitt, cognition, fiction. Eur Romant Rev 24:437–452

Syme K, Hagen EH (2020) Most anguish isn't an illness but an evolved response to adversity. Psyche, 29 Sept 2020, https://psyche.co/ideas/most-anguish-isnt-an-illness-but-an-evolved-response-to-adversity

Tinsley HEA, Tinsley DJ (1986) A theory of the attributes, benefits and causes of leisure experience. Leis Sci 8:1–45

Tolstoy L (1872) A confession. Unabridged Dover (2005) republication of the Aylmer Maude translation, as published by Oxford University Press, London, 1921

Wenner M (2009) Smile! It could make you happier: making an emotional face—or suppressing one—influences your feelings. Sci Am Mind Brain, Sept/Oct 2009, http://www.scientificamerican.com/article/smile-it-could-make-you-happier/

Wilde O (1891) The picture of Dorian Gray. Three Sirens Press, New York

Wilson TD, Reinhard DA, Westgate EC, Gilbert DT, Ellerbeck N, Hahn C, Brown CL, Shaked A (2014) Just think: the challenges of the disengaged mind. Science 345:75–77

Wisman A (2006) Digging in terror management theory: to use or lose the symbolic self. Psychol Inq 17:319–327

Chapter 10
Extension of Self

Here is the core of the enigma. This little consciousness, this feeling of a specific me, demands that it accompany us into infinity. (Maeterlinck 1913)[1]

Image courtesy of Christiane Beauregard (https://www.christianebeauregard.com/fr/portfolio/)

With 'discovery of self' (Chap. 4), *Homo sapiens* was launched on a trajectory of genetic and cultural evolution (Chaps. 5 and 6) unlike any other in the history of life. Building on primitive instincts for survival and reproduction inherited from

---

[1] Quoted from Choron (1964).

proto-humans (and shared with other animals), this evolutionary journey gave us adaptive motivations and behaviours—many of which are uniquely human—associated with Survival Drive (Chap. 8) and Sexual/Familial Drives (Chap. 7) (Box 9.1).

But somewhere along this journey we also evolved two additional, uniquely human drives—Leisure Drive (Chap. 9) and Legacy Drive (the present chapter). And with this we became a new species of sorts—one that I think could be appropriately named *Homo absurdus*—*human that spends its whole life trying to convince itself that its existence is not absurd* (Aarssen 2018, 2020). Absurdity is when there is '… a conspicuous discrepancy between pre-tension or aspiration and reality' (Nagel 1971). And as Albert Camus (1956) put it, 'Man is the only creature who refuses to be what he is'. The uniquely human capacity to foresee one's own eventual death (Chap. 4) normally evokes a well-spring of terror and denial (Choron 1964; Becker 1973; Solomon et al. 2015). An eloquent passage from the nineteenth-century English essayist, William Hazlitt (1827), provides a poignant description:

> To see the golden sun and the azure sky, the outstretched ocean, to walk upon the green earth, and to be lord of a thousand creatures, to look down giddy precipices or over distant flowery vales, to see the world spread out under one's finger in a map, to bring the stars near, to view the smallest insects in a microscope, to read history, and witness the revolutions of empires and the succession of generations, … to traverse desert wildernesses, to listen to the midnight choir, to visit lighted halls, or plunge into the dungeon's gloom, or sit in crowded theatres and see life itself mocked, to feel heat and cold, pleasure and pain, right and wrong, truth and falsehood, to study the works of art and refine the sense of beauty to agony, to worship fame and to dream of immortality, to have read Shakespeare and belong to the same species as Sir Isaac Newton; to be and to do all this, and then in a moment to be nothing, to have it all snatched from one like a juggler's ball or a phantasmagoria; there is something revolting and incredible to sense in the transition, and no wonder that, aided by youth and warm blood, and the flush of enthusiasm, the mind contrives for a long time to reject it with disdain and loathing as a monstrous and improbable fiction.

Anxiety about eventual mortality starts in childhood and extends cross-culturally. Numerous studies in social psychology involving 'terror management theory' (Solomon et al. 2015)—inspired by the work of Becker (1973)—have shown how death reminders commonly evoke a wide range of behaviours associated with 'world-view and esteem defense and striving' (Burke et al. 2010).

Self-deception is, of course, characteristically human (Dobzhansky 1967; Becker 1971; Smith 2007; Trivers 2011). As Poet T.S Eliot mused, '… humankind cannot bear very much reality' (Eliot 1943, No. 1 of *Four Quartets*). But where did this terror of eventual mortality come from, and why are humans so primed to feel its anguish? Three potential hypotheses are considered below, where 'eventual mortality' anxiety is interpreted—in terms of genetic fitness—as either maladaptive, neutral, or adaptive:

(i) *'Eventual mortality' anxiety is just ancient 'survival instinct' gone awry, misemployed by a fitness trade-off cost of time-awareness and personhood—i.e. possibly* **maladaptive** *in terms of genetic fitness.*

According to this hypothesis, time-awareness and personhood gives us knowledge of eventual mortality, which automatically deploys our primitive survival instinct, thus triggering its usual emotional response: anxiety. This anxiety possibly imposed a fitness cost for our ancestors—*but one that was worth paying, because the fitness benefits of time awareness and personhood were far greater* (Chap. 4). Survival instinct in other mammals is mostly about responding with 'fight, flight, or freeze' (accompanied by fear) to perception of a *looming* danger (e.g. attack from a predator or a rival)—or responding with frantic, fearful desperation to an *immediate* or impending shortage of an essential resource (e.g. starvation). Importantly, these mortality risks all involve physical pain (from injury or hunger), and the anxiety felt is adaptive because it evokes behaviour that minimizes these risks. The above responses of course also characterize expressions of traditional Survival Drive in humans (Chap. 8).

Here then is the crucial question: Is it possible that deployment of survival instinct in humans (because we can vividly imagine ourselves in the future) need not (as for other mammals) require an immediate or imminent threat to continued existence? In other words, did humans inherit a survival instinct so astute and over-powering that it also manifests as fear even in response to events that we know will only eventually happen, like death? The answer, according to this hypothesis, is yes—i.e. humans, and humans alone, have a survival instinct that is so acutely primed that it routinely compels us to be anxious about our own literal material death, *even though it can only be imagined, as an eventuality, sometime in the future, resulting (if young) even in the distant future from just ordinary old age, and even peacefully, without violence, injury or even pain—and even if (and while) all of this resides largely in the subconscious mind.*

If this is true, then the anxiety here is not adaptive because it does not evoke behaviours that minimize the mortality risk. Consequently, the Survival Drive at work here represents a kind of 'misfire'. Its emotional cost may or may not also have imposed a genetic fitness cost for our ancestors. But if it did (i.e. impose a cost that partially—but only partially—compromised the fitness benefit of time-awareness and personhood), then we might expect natural selection to have favoured cognitive domains like Leisure Drive (Chap. 9) that—through distractions—at least partially mitigated the anxiety.

(ii) *'Eventual mortality' anxiety is/was just a by-product of 'fear of the unknown'—possibly* **neutral** *in terms of genetic fitness.*

In this case, the anxiety had an emotional cost only, probably without imposing any significant fitness cost (or benefit). Imagining a non-violent and painless death (e.g. in old-age sleep, happening sometime eventually in our senior years) is just as much a mystery today as it was for our ancestors. We have no idea what the experience will be like, or even whether there will be anything to experience at all. Eventual mortality then can be considered as just one item in a list of several unknown and unseen things in the human experience that provide no clue about what they are—like quiet dark. Behind some of these, however, at least some of the time (but regularly enough for ancestors), danger was lurking—maybe a predator or an ambush by competitors. Accordingly, humans apparently evolved a general all-purpose, hard-wired, instinctual caution, and sometimes fear (despite its emotional cost), regarding anything and everything that couldn't be understood, couldn't be sensed, or couldn't be predicted, because this circumspection was, on average, good for gene transmission success in ancestors.

Importantly, however, this differs from the survival instinct triggered by *known* hazards (like lack of food, or an attacker that is in plain view, or that makes a familiar sound signalling its potential presence nearby). Much or most (and in some cases virtually all) of the time, when ancestors were confronted with a mysterious unknown, *there was really no danger 'lurking in the dark' at all*. Importantly, the latter, according to this hypothesis, was true all of the time with respect to 'eventual mortality' anxiety. In other words, a *general* anxiety ('just in case') about things unknown was undoubtedly adaptive for our ancestors, but the *specific* anxiety about one's imagined eventual death, at some unknown time in the future, *never was*, because 'eventual' death was never a 'danger'. And neither was the anxiety here maladaptive. As quoted in an earlier chapter, this is cleverly captured in the definition of life from Bierce (1906): 'LIFE, n. A spiritual pickle preserving the body from decay. We live in daily apprehension of its loss; yet when lost, it is not missed'.

If, however, anxiety about one's imagined eventual death, at some unknown time in the future, *was ancestrally maladaptive,* then (again) we might reasonably expect natural selection to have favoured cognitive domains—like Leisure Drive (Chap. 9)—serving to buffer the anxiety.

*(iii) 'Eventual mortality' anxiety was directly favoured by natural selection—i.e. **adaptive** in terms of genetic fitness.*

In this case, the anxiety *itself* directly promoted gene transmission success in ancestors. This seems counter-intuitive at first, but according to this hypothesis, the anxiety here is not really associated with the eventual experience of dying. Clearly our ancestors inherited strong motivations that prolonged survival, and so—since evolving mortality salience—have long been uneasy with the prospect of losing material life, but they essentially acquiesced to the inevitability of death. But according to this hypothesis, 'eventual mortality' anxiety is associated more directly with worry about what eventual death *imposes*: loss of *mental* life—i.e. impermanence of the 'inner self'. Hence, it is not really rooted in what is traditionally understood as a survival instinct, or 'survival drive'. Self-impermanence anxiety (introduced in Chap. 4) is about worrying that one's life is absurd—pointless, without purpose, meaningless—not primarily because time eventually brings an end to material life, but more specifically, because of the worry that *time inevitably annihilates all that we do, and all that we are.* For most modern human minds, the prospect of this is largely unimaginable, unacceptable, or categorically denied.

This curse of self-impermanence anxiety is the fundamental human predicament of *Homo absurdus*. Somehow, our ancestors evolved to become the only creature afflicted with unrelenting obsession to find 'purpose' and 'meaning' in terms of suppositions and hopes/beliefs about domains for extension of 'self' beyond mortal existence. But how could evolution produce such a predicament—such entrenched absurdity? How could natural selection shape a mind determined to '… believe passionately in the palpably not true … the chief occupation of mankind' (H.L. Mencken (1880–1956)—quoted from Pinker (1997)), and at the same time, plagued with perpetual uncertainty of its own alignment? A mind so easily deluded and riddled with angst would seem to have compromised ancestral reproductive success.

But mostly it didn't, according to this hypothesis, because the effects of self-impermanence anxiety are routinely mitigated by 'Legacy Drive'—an inclination to sense that the 'inner self' not only exists apart from the material body, but also (unlike the latter) *need not be impermanent*. It is a drive to leave (and an inclination to believe that one *can* leave—despite knowledge of inevitable mortality) something of oneself, a legacy, for the future. Legacy Drive then (examined below), in a way, 'comes to terms' with mortality salience. And yet, at the same time, it is always just a delusion. Consider that today, for every deceased human that has ever existed (save for a miniscule micro-fraction), it is as though they never did.

Only genes have legacy (Dawkins 1989). But Legacy Drive has a powerful hold on human motivation nonetheless, because (according to this hypothesis) it has evolutionary roots in an ancestral attraction to 'memetic legacy' through offspring: i.e. through feeling a sense that one can create a lasting 'carbon-copy' of selfhood by shaping the malleable minds (and hence the 'inner self') of offspring (and grand-offspring)—to instil within them the same (or at least some of the same) components (memes) of mental life (i.e. the thoughts, perceptions, inclinations, ideas, values, temperament, beliefs, attitudes, character, esteem, conscience, pride, ego, psyche, will, intellect, personality, hopes, fears, wishes, dreams, intentions, goals, aspirations, knowledge, skills, and virtues) that define who and what you think you are.

Our distant ancestors were undoubtedly clever enough to have surmised that the naivety of children and grandchildren makes them rather easily inspired and manipulated to conform to one's standard of expectation—i.e. to mirror one's own sense of 'self'—and thus to serve as a symbolic extension of oneself. This is also mostly a delusion, of course, with parents easily fooled (Harris 1998; Plomin 2019). But importantly, the reward nevertheless goes to gene transmission because offspring are the very vehicles of genetic legacy, including for genes that might inform Legacy Drive (as well as Leisure Drive) (Aarssen 2007). In other words, this affective response—fear and anguish from knowing there is no immortality, and worry that there is no self-permanence, combined with delusional belief in amelioration through achievement of memetic legacy—actually propelled copies of our ancestors' genes into us (Aarssen 2010). As Barash (2012) put it, 'Maybe awareness of mortality isn't merely a tangential consequence of consciousness but its primary adaptive value, if it has the effect of inducing people to seek yet another way of rebelling against mortality: by reproducing'.

Legacy Drive

The greatest use of life is to spend it on something that will outlast it. (William James (1842–1910), quoted from Perry (1935))

Self-impermanence anxiety buffers can manifest, therefore, not just as distractions (Chap. 9), but also in a distinctively different sense—as delusions for leaving recognition, remembrance, or some essence of one's personal identity for the future (Box 9.1) (Aarssen and Altman 2006; Aarssen 2007). Hence, whereas Leisure Drive facilitates 'meaning' in terms of hedonic wellbeing, as an 'escape from self' (Chap. 9), Legacy Drive facilitates meaning in terms of eudaimonic wellbeing, as an 'extension of self'. Their distinction is real and important; recent experimental studies have shown that a satisfying life can be hedonically happy but eudaimonically meaningless, whereas in other cases it can be eudaimonically meaningful but hedonically unhappy (Baumeister et al. 2013; Delle Fave et al. 2013).

Legacy Drive then serves to evoke a sense of 'meaning' for one's life—the self— that is defined entirely by deployment of an effective delusion: that time *need not* inevitably annihilate all that one does and all that one is. It doesn't matter that it's a delusion; it only matters that the delusion normally works (as do the distractions of Leisure Drive) in calming the troubled mind. And of course this is more likely if one does not discover the delusion (or can remain delusional about the delusion). But preferably, (as proposed in Chap. 12) it can also work if one can learn how to manage it effectively and responsibly, through understanding—both intellectually and compassionately—its deep evolutionary roots.

Biocultural evolution has given us a drive for legacy delusions that incorporate three distinct ventures for 'extension of self': *symbolic immortality* (through parenthood or accomplishment); *afterlife transcendence* (through religion/spiritualism); and a variety of cultures (including religion) that embody *belief/membership in something 'larger than self'* (Solomon et al. 2015). The latter involves 'fusion identity' (Swann and Buhrmester 2015)—a visceral sense of 'oneness' with a cultural group and its individual members, thus linking one's identity to the larger group identity (commonly proclaimed with symbolism, e.g. uniforms, rituals)—rooted in our ancestral fitness advantage from (and hence ingrained capacity for, and enjoyment derived from) socialization and cooperative alliances (Chap. 8). Importantly, because the group's identity (its essence/purpose/value/cultural worldview) is bigger than the individual—and because its existence started before (and is anticipated to continue after) one's individual existence—*a sense of oneness with the group provides an effective delusion for thinking that one's 'inner self' can endure beyond mortal existence*. Examples of larger-than-self, 'group identity' cultures include ethnicities, patriotisms, political parties/ideologies, career cultures (e.g. in education, law, science, engineering, medicine), military battalions, sports teams (memberships or fans), and of course, religions. Regarding the 'religious experience', William James (1904) pondered: 'The only thing that it unequivocally testifies to is that we can experience union with something larger than ourselves and in that union find our greatest peace'.

Religion

The human individual knows that he must die, but has thoughts larger than his fate. ...
Religion is an effort to be included in some domain larger and more permanent than mere
existence. (Feibleman 1963)

A sense of legacy from religion is, in one sense, associated of course with faith in doctrines (involving theism/deism/spirituality) that promise some kind of afterlife. Belief in a future life fulfils what Sigmund Freud (1928) recognized as '... the oldest, strongest and most insistent wish of mankind'. As discussed in Chap. 4, evidence from paleoanthropology strongly suggests that the imaginations of our ancestors may have been sufficiently creative for conjuring such superstitions and cultivating them in symbolisms and rituals dating from at least 50,000 years ago. Even many atheists are inclined to believe in afterlife and other supernatural ideas (Lawton 2019). In the words of Malinowski (1931), 'Religion ... can be shown to be intrinsically although indirectly connected with man's fundamental, that is, biological needs. Like magic it comes from the curse of forethought and imagination, which fall on man once he rises above brute animal nature'.

Religiosity remains conspicuous and powerful in modern culture, with many dozens of varieties to choose from—providing reassurance, for the faithful, that the 'self' (mental life) need not be impermanent, even while knowing that the body (material life) is. This is the so-called transcendent, or 'vertical' component of the fitness benefit of religion, both ancestrally and today—i.e. as a domain for legacy, through everlasting life of the 'soul'. As Kaufmann (1958) put it, 'Man is the ape that wants to be a god'. Even the most devoutly religious people know, however, that all religions are just delusions—except one, of course.

For our ancestors, as well as today, organized religion has also had an important 'horizontal' component: congregational affiliation. Worshipping memberships like churches, synagogues, mosques, and temples can provide at least three significant benefits for genetic fitness:

(i) By reinforcing one's confidence in the 'vertical' component (i.e. 'our God and his promises of salvation must be real if there are so many fellow believers').

(ii) As a vehicle for bolstering self-esteem (in terms of membership within a 'larger-than-self' cultural worldview), and a sense of memetic legacy (from attainment of social status/power through personal testimony before fellow parishioners, and personal accomplishment in the business of the religious institution).

(iii) By serving as an incentive to behave in ways that promote pro-social reciprocal exchange benefits of group membership—e.g. by not stealing, lying, murdering, etc.—because the threatened consequences of transgression involve not only shaming by the group against the perpetrator (and hence compromising one's intrinsic 'need to belong'), but also (supposedly) banishment of the soul to eternity in a bad place (e.g. hell). What was good for the prosperity of the social group was good also for the gene transmission success of resident members.

Our ancestors probably also enjoyed an additional, possibly more ancient, benefit from belief in the supernatural: answers (when no practical ones could be found) regarding the mysteries of life and nature—thus satisfying the restless human curiosity, and calming fears of the unknown. The answers here were of course interpreted in terms of favours, judgements, and interventions of a 'higher power', involving spiritualism and/or deity. And the early shamans, priests, prophets, and their esteemed disciples were also likely to have enjoyed elevated social status and greater attractiveness to potential mates. Fascination with the supernatural is reflected in modern cultural products like séances and ballet, aiming to 'defy the corporeal in a quest for the ethereal' (Campbell 2019)—and like magic and of course religion: 'You never see animals going through the absurd and often horrible fooleries of magic and religion. ... Only man behaves with such gratuitous folly. It is the price he has to pay for being intelligent but not, as yet, quite intelligent enough' (Huxley 1932).

Abundant evidence now indicates that attraction to religiosity has a partial genetic basis and that religious people generally have more children than non-believers (Rowthorn 2011). A predilection for superstitions then is evidently in our genes (Lawton 2017a). Religiosity, for those who 'believe' (and 'behave' accordingly), is not only an effective self-impermanence anxiety buffer (calming the intrinsic fear of failed legacy); it can calm general fears of the unknown and unexplained and promote social order and cohesion—group prosperity—and hence individual prosperity of resident members. Plus—because public dedication as a 'follower' normally evokes trust from other in-group members—religion bolsters one's local reputation, including with potential benefits through mate attraction. All of it delivered genetic fitness for ancestors.

Religiosity then is a fairly obvious product of biocultural evolution (Fuentes 2020). Interpretations of its evolutionary roots have been explored in no less than 28 recent books published in the span of just 17 years (Box 10.1). Some suggest that cultures of secularism and atheism might further their causes by learning from the 'playbook' of religiosity (Lawton 2017b). The old debates, therefore, between evolution and creationism (still active in some realms) are misguided; creationism is not in conflict with evolution—it is a product of it.

Box 10.1: Recent Books Exploring the Evolutionary Roots of Religion

- Boyer P (2001) *Religion Explained*. Basic Books, New York.
- Atran S (2002) *In Gods We Trust: The Evolutionary Landscape of Religion*. Oxford University Press Oxford.
- Wilson DS (2003) *Darwin's Cathedral: Evolution, Religion, and the Nature of Society*. University of Chicago Press, Chicago.
- Hamer D (2004) *The God Gene: How Faith is Hardwired Into Our Genes*. Doubleday, New York.
- Pyysiainen I (2004) *Magic, Miracles, and Religion: A Scientist's Perspective*. AltaMira Press, Walnut Creek, CA.
- Shermer M (2004) *The Science of Good and Evil: Why People Cheat, Gossip, Care, Share, and Follow the Golden Rule*. Henry Holt, New York.
- Kirkpatrick LA (2005) *Attachment, Evolution, and the Psychology of Religion*. The Guilford Press, New York.
- Wolpert L (2006) *Six Impossible Things Before Breakfast: The Evolutionary Origins of Belief*. Norton, New York.
- Dennett DC (2007) *Breaking the Spell: Religion as a Natural Phenomenon*. Penguin Viking, New York.
- Bulbulia J, Sosis R, Genet C, Genet R, Harris E, Wyman K, (Eds.) (2008) *The Evolution of Religion: Studies, Theories, and Critiques*. Collins Foundation Press, Santa Margarita, CA.
- Steadman LB, Palmer CT (2008) *The Supernatural and Natural Selection: The Evolution of Religion*. Paradigm, Boulder, CO.
- Feierman JR, ed. (2009) *The Biology of Religious Behavior: The Evolutionary Origins of Faith and Religion*. ABC-CLIO, LLC, Santa Barbara, CA.
- Wade N (2009) *The Faith Instinct: How Religion Evolved and Why It Endures*. Penguin, New York.
- Voland E, Schiefenhovel W (Eds.) (2009) *The Biological Evolution of Religious Mind and Behavior*. Springer, New York.
- Hinde RA (2010) Why Gods Persist: A Scientific Approach to Religion.
- Rossano MJ (2010) *Supernatural Selection: How Religion Evolved*. Oxford University Press, Oxford.
- Schloss J, Murray M (Eds.) (2010) *The Believing Primate: Scientific, Philosophical, and Theological Reflections on the Origin of Religion*. Oxford Univ. Press, Oxford.
- Stewart-Williams S (2010) *Darwin, God and The Meaning of Life: How Evolutionary Theory Undermines Everything You Thought You Knew*. Cambridge University Press, Cambridge.
- Wright R (2010) *The Evolution of God*. Little, Brown, New York.
- Bering J (2011) *The Belief Instinct: The Psychology of Souls, Destiny, and the Meaning of Life*. Norton & Co., New York.

(continued)

Box 10.1 (continued)

- Bellah RN (2011) Religion in Human Evolution: From the Paleolithic to the Axial Age. Belknapp Press, Cambridge.
- Shermer M (2011) The Believing Brain: From Ghosts and Gods to Politics and Conspiracies—How We Construct Beliefs and Reinforce Them as Truths. St. Martin's Press, New York.
- Watts F, Turner LP (2014) Evolution, Religion, and Cognitive Science: Critical and Constructive Essays. Oxford University Press, New York.
- Geertz AW (Ed.) (2014) Origins of Religion, Cognition and Culture. Routledge, New York.
- Wathey JC (2016) The Illusion of God's Presence: The Biological Origins of Spiritual Longing. Prometheus Books, New York.
- Torrey EF (2017) Evolving Brains, Emerging Gods: Early Humans and the Origins of Religion. Columbia University Press, New York.
- Asma ST (2018) Why We Need Religion. Oxford University Press, New York.
- Turner JH, Maryanski A, Klostergaard Petersen A, Geertz AW (2018) The Emergence and Evolution of Religion By Means of Natural Selection. Routledge, New York.

Parenthood

Man has a hope, perhaps an illusory one, that he somehow survives in his descendants. A life devoted to one's family and to one's progeny (biological or even adopted) seems to acquire a meaning; it may be experienced as capturing a particle of an immortality which is beyond the reach of an individual. (Dobzhansky 1967)

Several recent experimental studies of mortality priming have shown that the perception of parenthood can be an effective death-anxiety buffer (Wisman and Goldenberg 2005; Fritsche et al. 2007; Mathews and Sear 2008; Zhou et al. 2008, 2009; Yaakobi et al. 2014). On one level this might be interpreted as a consequence of traditional life history theory, applied broadly to more than just humans:

…fertility behaviour is likely to be particularly sensitive to mortality levels and patterns, given the key importance of mortality in determining the payoffs to life history decisions such as when to stop growing and reproducing, how to allocate investment between quantity and quality of children, etc. … High and unpredictable mortality regimes are likely to favour an early start to reproduction and high fertility, whereas low and stable mortality is predicted to lead to later and lower fertility. (Mathews and Sear 2008)

Alternatively (or in addition), as interpreted here, perceptions of offspring and parenthood provide deep-seated symbolism for immortality. Gene transmission per se of course is not a cognitive human goal (or at least never was prior to Gregor Mendel, less than two centuries ago). But pride in offspring (and also adopted children), forethought/planning for having them, and eager anticipation of their birth (or adoption) and family membership are all (we can reasonably presume) universal and uniquely human emotions and intentions. And they represent obvious manifestations of attraction to legacy, bolstering a delusional confidence in being able to leave extensions of 'self' (or at least the self that one aspires to, or at one time did) that might transcend death. As Schaller et al. (2010) recognize, an '…affective reward comes as we make progress toward the underlying evolutionary objective—when … the child scampers off the school bus and proudly produces a remarkable report card' (p. 336). Importantly here, it is the parent who especially feels pride (e.g. see Ashton-James et al. 2013; Brooks 2013; Cichy et al. 2013; Feiler 2013), and this is more than just a symptom of the primitive parental care instinct shared with other animals; it is a principal phenotype of the universal and uniquely human drive to leave something significant of oneself—a post-self—for the future. All other animals have offspring only as an incidental product of sex drive. Offspring for our species can of course also result for the same reason (Chap. 7). But only humans routinely hope and plan for having offspring, and then spend much of their lives seeking pride in them—and obsessively so in the popular culture of 'hyper-parenting' (Rosenfeld and Wise 2000).

The absurdity of this, as a dilemma for parents, however, is that they are normally set up for failure from the start. Their vision of legacy through meme transmission to offspring is largely just a delusion (Harris 1998; Plomin 2019). The children of the past who left the most descendants were not those dutifully committed to providing a meaningful life for their parents; instead they were those who boldly searched for and discovered their own personal domain for meaning within the cultures of their own generations, within which they needed to find success in mating—and incidentally, success for both memetic and genetic legacy, including for their parents.

Might this have had a powerfully important historical impact across generations, perpetually driving the replacement of old cultures by new ones? Perhaps the new ones didn't even need always to be better ones; they just needed to be different. Perhaps youth in general may be intrinsically attracted to them simply because, more often than not—for our ancestors in their youth—the new and different ones really were better, thus rewarding their reproductive success. Conceivably, biocultural evolution may have set up youth for merely the pursuit of new and contrasting cultures, not the literal attainment of better ones. 'It is good that the old should resist the young, and that the young should prod the old; out of this tension … comes a creative tensile strength, a stimulated development, a secret and basic unity and movement of the whole' (Durant and Durant 1968). After all, just as biological evolution is fueled by the blind, pitiless, indifferent effects of natural selection on genetic variation and novelty, cultural evolution is likewise fueled by similar effects of cultural selection on memetic variation and novelty (Box 6.4). From Wampole (2020): 'To argue that culture is degenerating might prove only one thing, namely that you haven't taken the time to look for the many (re)generative objects that are waiting there to be found, or that you don't quite know how to find meaning in the ones that displease you. We have a choice: either to judge the world as embittered critics who find little beauty in new things or to feel a tingle of elation at the seemingly infinite cultural improvisations that humans keep spinning out like silk'.

Nevertheless, Pulitzer Prize winning author Phyllis McGinley (1956) reasoned (over half a century ago) that: 'Women are the fulfilled sex. Through our children we are able to produce our own immortality, so we lack that divine restlessness which sends men charging off in pursuit of fortune or fame or an imagined Utopia'. Not surprisingly, therefore, traditionally men have been essentially in charge of defining and arranging—largely for themselves only—the domains and opportunities for both Leisure and Legacy through accomplishment (discussed below), as well as through religion (the 'imagined Utopia'); men have virtually always been in charge of religious institutions, and (throughout most of their history) the exclusive authors of their doctrines.

But an interesting cultural shift has developed in recent decades, as women (especially in developed countries) have gained considerable independence from their long history of patriarchal subjugation: growing numbers are embracing a 'childfree' lifestyle (Aarssen 2005; Shorto 2008; Sandler 2013; Kingston 2014), apparently abandoning the uniquely female opportunity for symbolic immortality through motherhood. They are now embracing a domain for legacy through accomplishment (see below) that was largely denied to their maternal ancestors. In terms of 'fulfilment', however, this Darwinian paradox may be understood more directly from the male perspective. In other words, it is men who, because of paternal uncertainty, have been the *unfulfilled* sex—accounting for their 'charging off' in pursuit of other forms of symbolic immortality. Indeed, some studies indicate that males, generally more than females, desire offspring following mortality priming (Wisman and Goldenberg 2005; Mathews and Sear 2008). As Sir Francis Bacon (1561–1626) put it:

> The perpetuity by generation is common to beasts; but memory, merit, and noble works are proper to men. And surely a man shall see the noblest works and foundations have proceeded from childless men, which have sought to express the images of their minds, where those of their bodies have failed. So the care of posterity is most in them who have no posterity. (Bacon 1985)

As discussed in Chap. 7, men—partly in order to minimize paternal uncertainty—have also taken control of female fertility throughout most of recorded history. Our unfulfilled male ancestors thus coerced female ancestors—many of whom happened not to have a particularly strong maternal instinct (or none at all)—to (nevertheless) bear offspring, including daughters who inherited their mothers' weak maternal instincts. This presents an intriguing evolutionary hypothesis explored further in the next chapter: women of the 'childfree' culture today, or some of them at least, may be descendants of these daughters (Aarssen 2005; Aarssen and Altman 2012).

Accomplishment

To a believer in resurrection and immortality, the temporal existence is merely a prepara-
tion for the existence to come; to those who hold death to be the ultimate dissolution, the
meaning of life must be found in what happens between birth and death, or else no meaning
can be found at all. A believer in immortality can therefore regard his personal salvation as
his ultimate concern, and the purpose of his life. If there is no immortality, the problem
becomes vastly more complex and difficult. A meaning of life can only be found in some-
thing greater than, but including, the personality. (Dobzhansky 1967)

Accomplishment commonly generates recognition or status, which may be earned
through dominance (usually involving coercion or intimidation) or through prestige
(from talents or deeds that evoke admiration, trust, and/or respect) (Henrich 2016).
In the cultural evolution of modern times, striving for personal accomplishment—
'personal branding'—has overshadowed both religion and parenthood as a domain
for Legacy Drive. Its pace started to pick up probably in the Renaissance, and was
bolstered by the Industrial Revolution. With the subsequent proliferation of capitalism and the politics of human rights and freedoms, and associated opportunities
grew exponentially—particularly in more developed countries, and especially for
women—for a wide array of human endeavours that normally draw attention and
earn acclaim or favourable reputation. Accomplishment today is fueled by attraction
to things like financial wealth and consumerism/materialism (e.g. Gountas et al.
2012; Shrum et al. 2013), successful business/institutional affiliations (e.g. Fox
et al. 2010), philanthropy (e.g. Wade-Benzoni et al. 2012), and many others (Maltby
2010): academic awards, competition for trophies/championships, volunteering for
community service, showing kindness to others, recognition of virtue or bravery/
heroism in a civilian crisis or in war, and pursuits of rewarding careers that commonly evoke admiration and/or prestige—e.g. in politics, government, law, teaching, social work, medicine, health care, the military, professional sports, research
and invention, and the arts (involving literature, acting, film production, musician-
ship, and other artistic displays and creative products) (Aarssen 2010). A nineteenth-
century essay from William Hazlitt (1805) is 'dripping' with Legacy Drive:

There are moments in the life of a solitary thinker which are to him what the evening of
some great victory is to the conqueror and hero—milder triumphs long remembered with
truer and deeper delight. And though the shouts of multitudes do not hail his success ... yet
shall he not want monuments and witnesses of his glory, ... that, as time passes by him with
unreturning wing, still awaken the consciousness of a spirit patient, indefatigable in the
search of truth, and the hope of surviving in the thoughts and minds of other men.

As Henry Wadsworth Longfellow (1885) mused: 'How can he be dead, who lives
immortal in the hearts of men?'

A sense of purpose/esteem/recognition then might be evoked by something as
common as praise or promotion from an employer for good work (Barrick et al.
2013), providing mentorship for a colleague, taking a stand in championing a
'cause' among peers within club memberships or public protest/grassroots movements, being a leader within a larger-than-self, 'group identity' culture (e.g. as a
favoured comrade of patriotism, or among fellow devotees to a particular sports

team or political party), commanding respect from rivals, or just earning popularity/ admiration from family, friends, or associates based on one's virtue of character (e.g. Hunter 2008; Greenwood et al. 2013). Celebrity/notoriety can, of course, also be earned from displays of non-conformity or misanthropy, ranging from harmless projections of 'cool' indifference, eccentricity or charisma, or pronouncements of rejection against fads/customs/societal norms (Warren and Campbell 2014), to deplorable extremes of deviant behaviour and criminal activity (Parnaby and Sacco 2004).

The 'cool' person is an intriguing case worth examining in more detail here. This is commonly regarded in modern culture as someone with a distinctive manner-ism—represented in certain ways of walking or speaking, or certain types of facial expressions or gestures (sometimes with accessories like sunglasses or cigarettes)— that evoke a demeanour of composure, self-confidence, and nonchalance towards situations where excitement or emotional vulnerability would normally be expected. This is usually also accompanied by a distinctive appearance or presentation—e.g. reflected in certain styles of dress or grooming (e.g. 'bed-head' hair, perpetual beard stubble), audacious choices for hobbies, or austere 'tastes' in music, preferred res-taurants, or home furnishings (e.g. exhibiting 'minimalism')—that evokes an impression of being impervious to the sway of fashions and conventions that are popular with 'ordinary' folks. These have been the trademarks of the traditional beatniks, bohemians, eclectics, freethinkers, eccentrics, dandies, dudes, hipsters, and elitists.

The intriguing question of course is: *why are we impressed with these things?* Is it because of our intrinsic attraction to anything that points to potential for defying the corporeal—an apparent indifference to mortality salience? Do people com-monly admire coolness and want to be like cool people because their characteristics signal, or suggest, a talent or constitution that could be deployed for dismissing or buffering the normally crippling private anxiety of mortality salience? Did our ancestors want to be around cool people (including as a potential mate) because they thought, or surmised subconsciously, that this talent might somehow rub off on them?

Perhaps coolness served our distant ancestors by 'announcing' truthfully (to potential adversaries and potential mates), an unflappable superiority: 'My talents for poise and emotional control have such high quality that nothing fazes me; I can handle any challenge with dispassionate ease'. The embedded message here is that this includes the challenge of responding effectively to the intrinsically human angst about self-impermanence. According to this hypothesis then, potential mates that were authentically cool were not only socially popular; they were sexually attractive and probably also perceived (correctly) as a good bet for being equipped to provide for (and to favourably inspire) one's offspring—all good things for gene transmis-sion success. Natural selection and cultural selection in our ancestral past, therefore, may have favoured (probably especially in males) dispositions for public manner-isms and styles of many sorts that projected a confident, self-determining, calm, and collective persona, with an elite 'knowingness'—all as truthful advertisements of these personifications. This would have also required of course the evolution of capacity to accurately detect and correctly interpret these advertisements in others.

But at the same time, the evolution of strategies for deception would have been inevitable. The latter, much of the time, can reward reproductive success just as well as the 'real deal'. Today therefore, the 'cool guy', much of the time, is likely to be a fake—a deliberate deception designed to attract the notice of others, serving only to bolster one's self-esteem (a handy tool-kit item, of course, for buffering self-impermanence anxiety). Like the mostly female culture of cosmetics, and the largely male culture of conspicuous wasteful consumerism, the culture of 'cool' is commonly also a false advertisement. In other words, the cosmetics user is often not really as young as she appears, the young male who buys expensive clothes or fast cars (that he can't afford) is often not really as rich as he appears, and the 'cool guy' whose aloof, countercultural style screams—'I'm too cool to care about convention or the latest trends that the masses follow'—is often just very concerned about appearing 'cool', and very talented in concealing that concern.

The latter is plainly evident in some men today whose clothing makes a state-ment—but not because it is flamboyant or expensive. Just the opposite; instead, it looks understated for the venue relative to the average or 'standard' expectation—e.g. because it looks (sometimes ever so subtly) like it is past its 'best-before' date, or because it is minimalist (e.g. a clean but plain undershirt, usually white, gray, or black, and untucked or half-tucked), or because it is a style from an earlier decade. On the surface of course, the 'cool guy' here—despite obviously being able to afford the latest fashion—looks like he just can't be bothered to make any signifi-cant effort in deciding what should be in his wardrobe. After all, only 'cool guys' can get away with that. But nothing could be further from the truth; the wardrobe was cleverly and carefully chosen (including even with intentional visits to the used clothing store)—*in order to give the appearance that it was not carefully chosen*. And so, just as with cosmetics and expensive fast cars, contrived 'coolness' will fool some of the people some of the time. But in many cases I suspect, the phoney cool are living with 'imposter syndrome'—a persistent fear of exposure as a fraud. The interesting and entertaining thing here lies in watching how these imposters need to continually reinvent themselves, as the masses discover and copy what is 'cool' (thus rendering it no longer cool)—and in trying to decipher who these many pre-tentious fakes are, lurking in our midst. It's easier of course to spot them, when you've been one yourself.

Conspectus

Maybe fictions and myths are not just errors to be dispelled, but productive illusions which allow us to thrive. … Life may be no more than a biological accident, and not even an acci-dent that was waiting to happen; but it has developed in us a random phenomenon known as the mind, which we can use to shield ourselves from the frightful knowledge of our own contingency. (Eagleton 2007)

The key argument in this chapter is that the many alternative engagements described above for securing a sense of legacy—i.e. recognition/status, belongingness, self-esteem, pride, transcendence, and fulfilment of one's perceived potential—are likely to be routinely successful in *fooling the mind/self into thinking that it need not be impermanent after all.* In other words, the salient human mind is predisposed (by evolutionary bequeathal) to be easily drawn into believing, naively, that it can have a 'post-self' by being larger than self—i.e. through association with a popular ideology, or 'membership' within a conspicuous cultural worldview, or by residing as a soul/spirit or as reincarnations throughout eternity (Solomon et al. 2010; Vail et al. 2010; Ellis and Wahab 2013); by attaining symbolic immortality as memes residing in the minds of one's offspring (Higginson and Aarssen 2011), or as fond remembrances or records of virtue/accomplishment/status/'coolness' residing in the minds of one's contemporaries (e.g. Hunter and Rowles 2005; Greenberg et al. 2010; Wojtkowiak and Rutjens 2011; Sligte et al. 2013; Perach and Wisman 2019), with a potential chain of influence passed on to their descendants in perpetuity; and/or as representations of legacy imbued in artefacts, inscribed on monuments, or stored in the written archives of history (Choron 1964; Shneidman 1973; Aarssen 2010; Cave 2012).

The remarkable trick of evolution here is not just that it makes us believe that memetic legacy brings meaning and that meaning is what we want; it also tricks us to keep trying by making us believe that there is meaning even in the trying—and even if frustrating. Accordingly, accounts of the long history of literature suggest that immortality is one of the most universal of human obsessions (Schellhorn 2008; Gollner 2013). Becker (1971) elaborates:

> *Modern man is denying his finitude with the same dedication as the ancient Egyptian pharaohs, but now whole masses are playing the game, and with a far richer armamentarium of techniques. The skyscaper buildings ... the houses with their imposing facades and immaculate lawns—what are these if not the modern equivalent of pyramids: a face to the world that announces: "I am not ephemeral, look what went into me, what represents me, what justifies me". The hushed hope is that someone who can do this will not die.*

These issues raise some interesting questions and speculations concerning the function of funeral ceremonies, and the usual human emotional responses to them. 'The cardinal fact is that all people everywhere take care of their dead in some fashion, while no animal does anything of the sort. ... Only a being who knows that he himself will die is likely to be really concerned about the death of others' (Dobzhansky 1967). Do funerals make us cry then because we will miss the deceased, or because funerals force us to confront the morbid fear of our own eventual demise? Do we embrace these rituals because it honours the life and memory of the deceased, or because this helps us to heal from the terrifying reminder that, like the deceased, 'my life is also impermanent'—and (like the culture of genealogy) helps to alleviate the resulting worry about: 'Will I be remembered? Will I leave something of my "self" for the future?' As Pinker (1997) put it, 'Ancestor worship must be an appealing idea to those who are about to become ancestors'. Also interesting is how the sadness and anxiety evoked by such direct confrontations with death are commonly alleviated by domains for Leisure Drive (Chap.

9)—through pleasurable distractions like going shopping, eating, and having sex (Mandel and Smeesters 2007; Schultz 2008; Birnbaum et al. 2011; Goldenberg 2013).

For two and a half millennia, since the dawn of philosophical speculation in ancient Greece, the meaning and significance of life and death have been among the principal problems of philosophy. Only the influential modern school of logical analysis, which prides itself on being rigorously "scientific", has declared these problems to be meaningless. There is, nevertheless, something irresistibly and overwhelmingly urgent and attractive about these particular "meaningless" problems. And although science cannot claim to solve them, it can perhaps furnish some information relevant to the speculations of the philosophers. (Dobzhansky 1967)

References

Aarssen LW (2005) Why is fertility lower in wealthier countries? The role of relaxed fertility-selection. Popul Dev Rev 31:113–126

Aarssen LW (2007) Some bold evolutionary predictions for the future of mating in humans. Oikos 116:1768–1778

Aarssen LW (2010) Darwinism and meaning. Biol Theory 5:296–311

Aarssen LW (2018) Meet *Homo absurdus*—the only creature that refuses to be what it is. Ideas Ecol Evol 11:90–95

Aarssen LW (2020) Meet *Homo absurdus*—the only creature that refuses to be what it is. Sci Animated, 5 Feb 2020. https://sciani.com/portfolio/meet-homo-absurdus-the-only-creature-that-refuses-to-be-what-it-is/

Aarssen LW, Altman ST (2006) Explaining below-replacement fertility and increasing childlessness in wealthy countries: legacy drive and the 'transmission competition' hypothesis. Evol Psychol 4:290–302

Aarssen LW, Altman T (2012) Fertility preference inversely related to 'legacy drive' in women, but not men: interpreting the evolutionary roots, and future, of the 'childfree' culture. Open Behav Sci J 6:37–43

Ashton-James CE, Kushlev K, Dunn EW (2013) Parents reap what they sow: child-centrism and parental well-being. Soc Psychol Personal Sci 4:635–642

Bacon F (1985) Francis Bacon: the essays. Penguin, London

Barash DP (2012) Homo mysterious: evolutionary puzzles of human nature. Oxford University Press, New York

Barrick MR, Mount MK, Li N (2013) The theory of purposeful work behavior: the role of personality, higher order goals, and job characteristics. Acad Manag Rev 38:132–153

Baumeister RF, Vohs KD, Aaker JL, Garbinsky EN (2013) Some key differences between a happy life and a meaningful life. J Posit Psychol 8:505–516

Becker E (1971) The birth and death of meaning: an interdisciplinary perspective on the problem of man, 2nd edn. The Free Press, New York

Becker E (1973) The denial of death. Simon and Schuster, New York

Bierce A (1906) The cynic's word book. Arthur F. Bird, London. Also published as *The Devil's Dictionary*. http://www.thedevilsdictionary.com

Birnbaum G, Hirschberger G, Goldenberg J (2011) Desire in the face of death: terror management, attachment, and sexual motivation. Pers Relat 18:1–19

Brooks R (2013) My child has achieved more than your child. Kidsinthehouse, 18 July 2013, https://www.kidsinthehouse.com/all-parents/parenting/my-child-has-achieved-more-than-your-child

Burke BL, Martens A, Faucher EH (2010) Two decades of terror management theory: a meta-analysis of mortality salience research. Personal Soc Psychol Rev 14:155–195

Campbell O (2019) How ballerinas defy the corporeal in a quest for the ethereal. Aeon, 20 May 2019, https://aeon.co/ideas/how-ballerinas-defy-the-corporeal-in-a-quest-for-the-ethereal

Camus A (1956) The rebel. Alfred A. Knopf, New York

Cave S (2012) Immortality: the quest to live forever and how it drives civilization. Crown Publishers, New York

Choron J (1964) Modern man and mortality. The Macmillan Company, New York

Cichy KE, Lefkowitz ES, Davis EM, Fingerman KL (2013) "You are such a disappointment!": negative emotions and parents' perceptions of adult children's lack of success. J Gerontol B Psychol Sci Soc Sci 68:893–901

Dawkins R (1989) The selfish gene, Rev. ed. Oxford University Press, Oxford

Delle Fave A, Brdar I, Wissing MP, Vella-Brodrick DA (2013) Sources and motives for personal meaning in adulthood. J Posit Psychol 8:517–529

Dobzhansky T (1967) The biology of ultimate concern. The New American Library, New York

Durant W, Durant A (1968) The lessons of history. Simon & Schuster, New York

Eagleton T (2007) The meaning of life. Oxford University Press, Oxford

Eliot TS (1943) Four quartets. Harcourt, San Diego

Ellis L, Wahab EA (2013) Religiosity and fear of death: a theory-oriented review of the empirical literature. Rev Relig Res 55:149–189

Feibleman JK (1963) Mankind behaving: human needs and material culture. Charles C Thomas, Springfield

Feiler B (2013) A truce in the bragging wars. NY Times, published 1 Feb 2013, http://www.nytimes.com/2013/02/03/fashion/time-for-a-truce-in-the-bragging-wars.html?pagewanted=all&_r=0

Fox M, Plunkett Tost L, Wade-Benzoni KA (2010) The legacy motive: a catalyst for sustainable decision making in organizations. Bus Ethics Q 20:153–185

Freud S (1928) The future of an illusion. Hogarth Press, London

Fritsche I, Jonas E, Fischer P, Koranyi N, Berger N, Fleischmann B (2007) Mortality salience and the desire for offspring. J Exp Soc Psychol 43:753–762

Fuentes A (2020) How did belief evolve? Sapiens, 26 Feb 2020, https://www.sapiens.org/biology/religion-origins/

Goldenberg JL (2013) Immortal objects: the objectification of women as terror management. Objectification and (De)humanization 60:73–95

Gollner AL (2013) The book of immortality: the science, belief, and magic behind living forever. Scribner, New York

Gountas J, Gountas S, Reeves RA, Moran L (2012) Desire for fame: scale development and association with personal goals and aspirations. Psychol Mark 29:680–689

Greenberg J, Kosloff S, Solomon S, Cohen F, Landau M (2010) Toward understanding the fame game: the effect of mortality salience on the appeal of fame. Self Identity 9:1–18

Greenwood D, Long CR, Dal Cin S (2013) Fame and the social self: the need to belong, narcissism, and relatedness predict the appeal of fame. Personal Individ Differ 55:490–495

Harris JR (1998) The nurture assumption: why children turn out the way they do. Simon & Schuster, New York

Hazlitt W (1805) An essay on the principles of human action. J Johnson, St. Paul's Churchyard, London. https://archive.org/details/anessayonprinci00hazlgoog

Hazlitt W (1827) On the feeling of immortality in youth. First published in the Monthly Magazine, March, 1827. Reproduced in Howe PP (ed.) (1934) The complete works of William Hazlitt, vol. XVII. J.M. Dent & Sons, London

Henrich J (2016) The secret of our success: how culture is driving human evolution, domesticating our species, and making us smarter. Princeton University Press, Princeton

Higginson MT, Aarssen LW (2011) Gender bias in offspring preference: sons still a higher priority, but only in men—women prefer daughters. Open Anthropol J 4:60–65

Hunter EG (2008) Beyond death: inheriting the past and giving to the future, transmitting the legacy of one's self. Omega 56:313–329

Hunter EG, Rowles DG (2005) Leaving a legacy: toward a typology. J Aging Stud 19:327–347

Huxley A (1932) Texts and pretexts: an anthology with commentaries. Chatto and Windus, London

James W (1904/1994). The varieties of religious experience. Random House Modern Library Edition, New York

Kaufmann W (1958) Critique of religion and philosophy. Harper and Row, New York

Kingston A (2014) The no-baby boom: social infertility, baby regret and what it means that shocking numbers of women aren't having children. MacLean's Mag, March 2014, http://www.macleans.ca/society/the-no-baby-boom/

Lawton G (2017a) Effortless thinking: the god-shaped hole in your brain. New Sci, 13 Dec 2017, https://www.newscientist.com/article/mg23631561-000-effortless-thinking-the-godshaped-hole-in-your-brain/

Lawton G (2017b) Faith of the faithless: is atheism just another religion? New Sci, 11 Apr 2017, https://www.newscientist.com/article/mg23431212-800-faith-of-the-faithless-is-atheism-just-another-religion/

Lawton G (2019) Why almost everyone believes in an afterlife – even atheists. New Sci, 20 Nov 2019, https://www.newscientist.com/article/mg24432570-500-why-almost-everyone-believes-in-an-afterlife-even-atheists/

Longfellow HW (1885) The poetical works of Henry Wadsworth Longfellow. Houghton, Mifflin and Company, Boston

Maeterlinck M (1913) La Mort. Bibliothèque-Charpentier, Paris

Malinowski B (1931) The role of magic and religion. In: Lessa WA, Vogt EZ (eds) Reader in comparative religion. Row Peterson, Evanston

Maltby J (2010) An interest in fame: confirming the measurement and empirical conceptualization of fame interest. Br J Psychol 101:411–432

Mandel N, Smeesters D (2007) Shop 'til you drop: the effect of mortality salience on consumption quantity. Adv Consum Res 34:600–601

Mathews P, Sear R (2008) Life after death: an investigation into how mortality perceptions influence fertility preferences using evidence from an internet based experiment. J Evol Psychol 6:155–172

McGinley P (1956) Women are wonderful: they like each other for all the sound, sturdy virtues that men do not have. Life Mag, 24 Dec 1956

Nagel T (1971) The absurd. J Philos 68:716–727

Parnaby PK, Sacco VF (2004) Fame and strain: the contributions of mertonian deviance theory to an understanding of the relationship between celebrity and deviant behavior. Deviant Behav 25:1–26

Perach R, Wisman A (2019) Can creativity beat death? A review and evidence on the existential anxiety buffering functions of creative achievement. J Creat Behav 53:193–210

Perry RB (1935) The thought and character of William James, as revealed in unpublished correspondence and notes, volume 1. Little Brown, New York

Pinker S (1997) How the mind works. Norton, New York

Plomin R (2019) The parenting myth: how kids are raised matters less than you think. New Sci, 22 May 2019, https://www.newscientist.com/article/mg24232310-800-the-parenting-myth-how-kids-are-raised-matters-less-than-you-think/

Rosenfeld A, Wise N (2000) The over-scheduled child: avoiding the hyper-parenting trap. St. Martin's Griffin, New York

Rowthorn R (2011) Religion, fertility and genes: a dual inheritance model. Proc R Soc B 278:2519–2527

Sandler L (2013) Having it all without having children. Time Mag, August 2013. http://time.com/241/having-it-all-without-having-children/

Schaller M, Neuberg SL, Griskevicius V, Kenrick DT (2010) Pyramid power: a reply to commentaries. Perspect Psychol Sci 5:335–337

Schellhorn GC (2008) Man's quest for immortality. Horus House Press, Madison

Schultz N (2008) Morbid thoughts make us reach for the cookie jar. New Sci 198:12

Shneidman ES (1973) Deaths of man. Quadrangle/New York Times Book Co., Oxford

Shorto R (2008) No babies? N Y Times, June 2008, http://www.nytimes.com/2008/06/29/magazine/29Birth-t.html?_r=2&ref=magazine&pagewanted=all&

Shrum LJ, Wong N, Arif F, Chugani SK, Gunz A, Lowrey TM, Nairn A, Pandelaere M, Ross SM, Ruvio A, Scott K, Sundie J (2013) Reconceptualizing materialism as identity goal pursuits: functions, processes, and consequences. J Bus Res 66:1179–1185

Sligte DJ, Nijstad BA, De Dreu CKW (2013) Leaving a legacy neutralizes negative effects of death anxiety on creativity. Personal Soc Psychol Bull 39:1152–1163

Smith DL (2007) The most dangerous animal: human nature and the origins of war. St. Martin's Press, New York

Solomon S, Greenberg J, Pyszczynski T, Cohen F, Ogilvie DM (2010) Teach these souls to fly: supernatural as human adaptation. In: Schaller M, Norenzayan A, Heine SJ, Yamagishi T, Kameda T (eds) Evolution, culture and the human mind. Psychology Press, New York, pp 99–118

Solomon S, Greenberg J, Pyszczynski T (2015) The worm at the core: on the role of death in life. Random House, New York

Swann WB, Buhrmester MD (2015) Identity fusion. Curr Dir Psychol Sci 24:52–57

Trivers R (2011) The folly of fools: the logic of deceit and self-deception in human life. Basic Books, New York

Vail KE, Rothschild ZK, Weise DR, Solomon S, Pyszczynski T, Greenberg J (2010) A terror management analysis of the psychological functions of religion. Personal Soc Psychol Rev 14:84–94

Wade-Benzoni KA, Plunkett Tost L, Hernandez M, Larrick RP (2012) It's only a matter of time: death, legacies, and intergenerational decisions. Psychol Sci 23:704–709

Wampole C (2020) Can culture degenerate? Aeon, 5 Aug 2021, https://aeon.co/essays/the-idea-of-cultural-degeneration-has-an-unsavoury-pedigree

Warren C, Campbell MC (2014) What makes things cool? How autonomy influences perceived coolness. J Consum Res 41:543–563

Wisman A, Goldenberg J (2005) From grave to the cradle: evidence that mortality salience engenders a desire for offspring. J Pers Soc Psychol 89:46–61

Wojtkowiak J, Rutjens BT (2011) The postself and terror management theory: reflecting on after death identity buffers existential threat. Int J Psychol Relig 21:137–144

Yaakobi E, Mikulincer M, Shaver PR (2014) Parenthood as a terror management mechanism: the moderating role of attachment orientations. Personal Soc Psychol Bull 40:762–774

Zhou X, Liu J, Chen C, Yu Z (2008) Do children transcend death? An examination of the terror management function of offspring. Scand J Psychol 49:413–418

Zhou X, Lei Q, Marley SC, Chen J (2009) Existential function of babies: babies as a buffer of death-related anxiety. Asian J Soc Psychol 12:40–46

Chapter 11
The Big Four Human Drives

We do not know exactly how much genetic change has taken place in mankind at different stages of its history. Modern man might or might not be able to survive, even if properly trained, in the environments of his ancestors of 100,000, or even 10,000 years ago. Or, if he survived, he might not be as efficient or as happy in those environments as his ancestors were. We do not know for sure. Neanderthal man may or may not have been capable of becoming a reasonably well adjusted citizen if raised in New York or in New Orleans. Perhaps some Neanderthals might have been fit to become Ph.D.'s and to be elected members of the Society of Sigma Xi. But, on the other hand, they may have been unfit for any now existing education. Modern women are alleged to experience greater difficulties in childbirth than did their great-grandmothers.[1] All this is conjectural and not rigorously proved. It is, however, a fallacy to assert that what is unproved did not occur. In point of fact, some of the above changes probably did occur. (Dobzhansky 1961)

Image courtesy of Zakariae BenKoudad (https://www.flickr.com/photos/izak_ben/5703940310)

---

[1] For example, see Walsh JA (2008) Evolution and the caesarean section rate. American Biology Teacher 70: 401–404. DOI: 10.2307/30163311.

© The Author(s), under exclusive license to Springer Nature Switzerland AG 2022
L. Aarssen, *What We Are: The Evolutionary Roots of Our Future*,
https://doi.org/10.1007/978-3-031-05879-0_11

In this chapter, we integrate themes from the last four chapters concerning the four major drives that define the pyramid of human needs (Box 9.1). As in the Kenrick et al. (2010) version (Box 8.1), the four-drives pyramid also assumes a developmental but integrative hierarchy, and this is signified by the arrow within the pyramid connecting across all levels. In other words, for the same reasons outlined by Kenrick et al. (2010), and echoing Maslow (1943), higher order goals/drives are generally more active at later developmental stages/ages, and are generally less likely to be satisfied if lower order needs are unmet. Lower level drives, however, can be activated at any stage (i.e. they are not replaced by higher level drives), and once developed, the activation of a drive, or 'goal system', will usually be triggered when relevant environmental cues are salient (Kenrick et al. 2010). Becker and Kenrick (2014) elaborate:

> Certain stimuli elicit stronger reactions than others, because they have more significant and/or consistent consequences in the ancestral (or developmental) past. Cognitive systems have thus evolved (or are biologically prepared to learn) a vigilance for stimuli relevant to fundamental goals. Neither the stimuli nor the goals exist in isolation; the psychological system has coevolved with features of the ecology.

Subselves

This speaks to the appealing notion of different 'subselves' (Martindale 1980; Becker and Kenrick 2014), defined by domain-level 'pyramid' goals (Box 11.1), activated by environmental cues [and underlying a central theme within two recent popular books (Kenrick 2011; Kenrick and Griskevicius 2013)]. Accordingly, for someone raised in the slums of Delhi, we can generally expect the 'Survival-drive-subself' to be engaged more vigilantly than for someone living in urban Tokyo, where we would expect to see more deployment of the 'Leisure-Drive-subself'. Similarly, we might expect activation of the 'Legacy-Drive-subself' versus the 'Leisure-Drive-subself' to be contingent on local ecology/culture. One recent study of responses to mortality salience provides an intriguing example of this: European Americans tended to choose responses that focused on achieving symbolic immortality (Legacy), while East Asians generally chose responses aimed at engaging in and enjoying life (Leisure) (Ma-Kellams and Blascovich 2012).

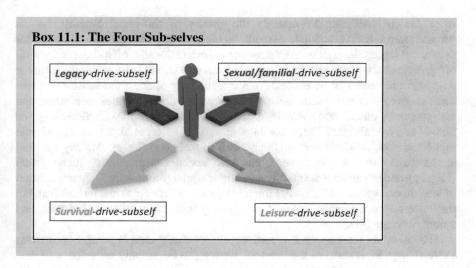

Box 11.1: The Four Sub-selves

Legacy-drive-subself *Sexual/familial-drive-subself*

Survival-drive-subself *Leisure-drive-subself*

The arrowhead in Box 9.1, 'collecting' the impact of all of the drives, resides in the pyramid apex, representing the ultimate but imperceptible evolutionary 'goal'—copying and transmission success for one's resident genes—subserved by the cognitive goals/drives of one's conscious (and/or subconscious) mind. And most importantly, regardless of rudimental need fulfilments from deployment of lower drives, the ultimate evolutionary goal remains unattained if there is no sex/mating, and may also be missed (even with sex/mating) if parenting is neglected—although there may be some effective compensation (inclusive fitness) if there is kin-helping.

As in the Kenrick et al. (2010) model, certain specific motivations in the four-drives pyramid may be deployed in solving problems across domains. For example, particular sources of enjoyment, providing the self-impermanence anxiety buffers of Leisure Drive, may cut across multiple levels, e.g. involving physiological needs (eating tasty foods), but also mating needs (sexual arousal). Attraction to religion/spiritualism/mysticism, career achievements, and showing kindness to others, all represent not just ventures for Legacy Drive; they also feel good (satisfying Leisure Drive), they garner resources and/or may earn favour within one's social group (thus reaping advantage for Survival Drive), and a reputation of success in these pursuits can also be attractive to potential mates (addressing Sexual/Familial Drives). Similarly, intimacy of course addresses sexual/familial drive, but it can also be an 'escape from self' distraction (Leisure) through the pleasure of sexual love, as well as an 'extension of self' delusion (Legacy) through the sense of having a 'soul mate', where 'two become one', and each then belongs to something 'larger than self'. Traditional Buddhist mindfulness meditation also instils a sense of belonging to something 'larger than self'—and at the same time, meditation in the 'modern view' focuses the mind on very specific soothing thoughts, blocking out troubling ruminations (Condon and Makransky 2020)—hence therapeutically releasing it from anxiety, including importantly the anxiety of self-impermanence, i.e. an 'escape from self'. Attraction to parenthood (and grand-parenthood)—an option for delusional legacy through meme uploading to impressionable offspring (and grand-offspring) minds—may also be triggered by Leisure Drive. In other words, intrinsic enjoyment rewards can be evoked from intentionally seeking the sense of attach-ment security and self-worth routinely connected with feelings of admiration and acceptance by others, and normally associated with close family relations (e.g. see Shaver and Mikulincer 2012; Yaakobi et al. 2014; Nelson et al. 2014). The simple joys from wonder, discovery, and creativity also feel good (Leisure) at any age, and at the same time, if conspicuous, can earn acclaim within one's social group (Legacy), and advertise mate quality as a fitness signal (Sex; Chap. 7). Accumulation of wealth, and spending it in shopping (consumerism) of course ensure survival, but they also advance relative status (Legacy), buy toys and vacations (Leisure), and attract romance (Sex).

The culture of team sports is a particularly interesting case in point. For Survival Drive (Chap. 8), it satisfies the need to 'belong', to be socially accepted, and thus promotes a sense of having reciprocal exchange benefits available from in-group ('us') membership (and successful competition against 'them', symbolized by the opposing team and its fans). Team sports also represent a venue for courtship display associated with Sex/Familial Drives, especially for males. Displays of physical strength and skill—especially involving personal risk—was a fitness signal for our female predecessors during mate choice (correlating with male ability to protect and provide), and so rewarded the reproductive success of our ancestors. Male sports fans therefore have an intrinsic attraction to (and curiosity about) such displays in other males as potential sexual rivals (Chap. 7). In addition, the need—as a sports team member and fan—to be a part of something 'larger-than-self', plus the drive to 'win' games of all kinds, may represent a delusional substitute for legacy through other domains (e.g. career accomplishments), of particular interest to males, histori-cally because of uncertainty of paternity. Legacy Drive (Chap. 10) engenders a notion of 'struggle' and 'challenge' and goal-directedness, and sport allows one to be engaged in that sense of striving for a goal, and so to satisfy that inner yearning to achieve something significant (before death)—even though it might only be mem-bership on the winning team at a pick-up hockey game played in an empty arena. Finally, team sport is of course pure entertainment—possibly one of the most potent, and oldest, cultural domains for distraction/escape from self (Leisure Drive; Chap. 9).

Physical fitness exercise/training, and associated club memberships, also pro-vides fitness signal displays (appealing to the Sex-Drive-subself), and at the same time they promote staying in 'good shape' (the Survival-Drive-subself), advertise-ment of 'personal branding', status competition, and feelings of belonging to some-thing 'larger than self' (the Legacy-Drive-subself), and regular endorphin/endocannabinoid rushes, e.g. the 'runner's high' (appealing to the Leisure-Drive subself). All of the big four human drives then are vividly represented in team sports, in physical fitness, and also in consumerism. No wonder that these three domains are among the most pervasive modern products of biocultural evolution.

Blending Legacy and Leisure Drives

In many cases, there is likely to be a blurred distinction, or blending, in the deployment of Legacy and Leisure Drives. Both, after all, represent explicit remedies in striving for an untroubled mind, and so, when these are working together, self-impermanence anxiety buffering is likely to be especially effective. Human achievements and triumphs define the history of cultural evolution (Chap. 5), in large part because they generously rewarded the reproductive success of ancestors. It should be no surprise, therefore, that we are routinely more than content, instinctively so (through evolutionary bequeathal), to endure the striving and struggling needed to reach our individually prescribed goals and achievements—even finding pleasure from the toil itself. As Csikszentmihalyi and LeFevre (1989) describe it: 'When both challenges and skills are high, the person is not only enjoying the moment, but is also stretching his or her capabilities with the likelihood of learning new skills and increasing self-esteem and personal complexity'. This is famously reflected in the depiction of *Sisyphus* from French author and philosopher, Albert Camus (1942):

> ... *At the very end of his long effort measured by skyless space and time without depth, the purpose is achieved. Then Sisyphus watches the stone rush down in a few moments toward that lower world whence he will have to push it up again toward the summit. He goes back down to the plain. It is during that return, that pause, that Sisyphus interests me. ... Each atom of that stone, each mineral flake of that night filled mountain, in itself forms a world. The struggle itself toward the heights is enough to fill a man's heart. One must imagine Sisyphus happy.*

Working hard at doing something that seems 'purposeful', including because it just 'needs to be done', may commonly therefore evoke a sense of leaving something significant about oneself for the future (Taggart 2017). But it can also just feel good, I suggest, simply because it requires keeping busy, hence arresting the mind firmly in the immediate present, thus providing diversion from self-impermanence anxiety. Accordingly, delusions of legacy and distractions of leisure may commonly be deployed at the same time in making meaning/happiness for one's life, all while remaining largely (and safely) oblivious to the fact that time will annihilate all that we do, and all that we are. As Eric Hoffer (1976) put it, 'A busy life is the nearest thing to a purposeful life'. In other words, if you can't keep calm, just keep busy (Box 11.2). And looking conspicuously *very* busy has even become a status symbol in modern culture (Bellezza et al. 2017). In my own world of academics, motivation fairly obviously involves a blend of leisure and legacy drives. Philosopher Zena Hitz (2020) describes this in recounting the early years of her academic career: 'The twin plants of intellectual joy ["lost in thought"] and of achievement in prestige and status had grown together so closely that, to me, their blossoming was indistinguishable, each from the other'.

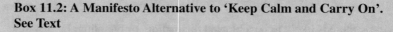

**Box 11.2: A Manifesto Alternative to 'Keep Calm and Carry On'.
See Text**

Another, perhaps more noble example of this blend is found in the sense of 'meaning' wrapped up in the purposeful work (during one's free-time indulgence) of helping others who are in need—thus providing a good old-fashioned dose of 'feel-good' (Leisure), and at the same time delivering a sense of having accomplished something likely to be valued and remembered by others (Legacy). Choron (1964) describes a view of this attributed to the famous Russian novelist Leo Tolstoy (1828–1910): '… life devoted to others is the only true "immortal" life in the sense that although it may be destroyed sooner than the selfish life, death cannot render it meaningless, as it does the other'.

The frenetic digital culture of social media and the Internet is perhaps the best example of recent biocultural evolution fueled by human obsession with chasing both legacy and leisure. Having more Facebook friends and Twitter followers can signal greater social status and bolster self-esteem, and perpetual tweeting and cell phone texting and emailing can provide on-demand reassurance of one's membership and recognition within a 'larger-than-self' social group (e.g. see Lenhart et al. 2015). And the allure of access to endless curiosities and entertainments on the World Wide Web provides free and instantaneous 'escape from self'. As Updike (2000) put it:

> Millions find a bliss of sorts in losing themselves in the vastness of the Internet, a phantom creation that sublimates the bulky, dust-gathering contents of libraries and supermarkets into something impalpable and instantaneous. The Web is conjured like the genie of legend with a few strokes of the fingers, opening with a phrase or two a labyrinth littered with trash, pitted with chat rooms, at once fascinatingly extensive and intensely private. Communication is antiseptically cleansed of all the germs and awkwardness of even the most mannerly transaction with another human body.

Another example of this blend can be found in recent pronatalist movements that involve attraction to large family size—typically supported by wealth, and combined often with religion (e.g. the 'Quiverfull' movement; Joyce 2010), but not in all cases (Brooks 2004; Kaufmann 2010; Rowthorn 2011; Caplan 2012). It is interesting to consider whether this represents a distinct motivational 'sub-domain' that has perhaps never (or only locally or occasionally) had opportunity, through evolutionary bequeathal, to be conspicuous within human populations—'parenting drive' (Aarssen 2007). This is not the same thing as attraction to legacy through parenthood, or to pleasurable rewards (feelings of self-worth and admiration by others) that may be triggered by it (as discussed above). 'Parenting drive' here is defined as attraction to memetic legacy through influence on offspring, but one that is also heavily infused with intrinsic attraction to a particular kind of enjoyment reward at the same time—triggered specifically (odd as it may seem to some) by the hard work of parenting. Research has shown that, as in Camus' (1942) *Myth of Sisyphus*, there is often a sense 'meaning' in purposeful toil and mundane routine (Baumeister et al. 2013). And, as discussed in Chap. 9, pre-occupation with hard work (especially when purposeful) leaves little time for worry or depressing thoughts, including those rooted in self-impermanence anxiety. A distracting pleasure then—like that of leisure—is obtained from just staying busy. And the 'busy work' of parenting is available in greater abundance, of course, with increasing family size (Angeles 2010; Nelson et al. 2013).

Important to note here is that weak parenting drive—despite its obvious disadvantage for evolutionary fitness—probably never had widespread opportunity to be strongly disfavoured by natural selection. Historically, many or most women were essentially forced, by patriarchal subjugation and/or religious imperatives (also controlled by men), to bear offspring (often many) regardless of whether they had any intrinsic desire to be hard-working mothers (and presumably, many didn't). Like homosexuality (Chap. 7), therefore, the 'Darwinian paradox' of weak parenting drive—conspicuous today in the now popular 'childfree' culture (see Chap. 10)— may be another example of the 'failed disfavouring selection' hypothesis; i.e. some contemporary behaviours, rather than a consequence of being favoured by natural selection in the ancestral past, may instead just be a consequence of *not having been disfavoured* (Aarssen 2005). Failed disfavouring selection resulting from historical patriarchal subjugation may also account for two additional Darwinian paradoxes appearing more recently in popular culture: a decreasing interest in heterosexual partnerships (the 'freemale' culture) (Trimberger 2005; Davies 2008; Zeidler 2018), and a decreasing interest in sex (with partners) altogether—the 'sex recession' culture (Julian 2018).

With the relatively recent erosion of patriarchy, however, many women can now make their own choices in virtually every aspect of their lives, including their intrinsic drives for escape from self and extension of self. And dozens of recently published books extol the gratifications of childfree living, virtually all of which involve maximizing opportunity for leisure and personal life goals/achievements (legacy), mostly for women. But choosing to be 'childfree'—as women are increasingly free to do—means zero gene transmission through direct lineage. Accordingly, selection against weak parenting drive may soon be ramping up (Aarssen 2007). If so, we might ask whether this selection could, within say a generation or two, spell an abrupt end to the now popular 'childfree' culture that accounts in part for the population implosion (below-replacement fertility) that has surfaced in many developed countries in recent years (Aarssen 2005; Aarssen and Altman 2006, 2012). The same fate might be predicted for the 'free-male' and 'sex-recession' cultures. As discussed in Chap. 6, any conspicuous human motivation or popular culture, informed by socially transmitted inheritance, that compromises gene transmission success (a 'Darwinian Paradox'), is likely to be short-lived. Important potential implications of this will be explored in the next chapter.

Being fooled by the soothing delusions of 'post-self' legacy, and distracted by the lure of pleasurable, 'outside-of-self' leisure—and hence also their inducements by mortality awareness and anxiety—all turn out then to be in the best interests of resident genes. And these interests are served only if being fooled and distracted can be sufficiently maintained (often enough, and long enough) until reproductive maturity is reached, thus effecting potential for successful gene transmission to descendants. As age advances beyond reproductive maturity, however, one may become less easily fooled and distracted. Legacy Drive, it seems then, presents as a kind of revolt—a confident denial—against self-impermanence, with perhaps (as an empirical prediction) greater activation expected prior to mid-life—whereas Leisure Drive may serve as more of a therapeutic reconciliation, especially perhaps in later life, when one may be more likely, in some respects, to accept the inevitability of self-impermanence. Propensity for self-deception then equips us with more than just skill for deceiving others (Trivers 2011); it protects us from knowing ourselves too much for our own good—or more precisely, too much in terms of compromising transmission success for resident genes. It convinces us, periodically, and for a while, that our existence is not absurd.

As considered above, male displays of accomplishment/fame/status (in seeking Legacy) and displays of intellectual, artistic, and athletic skills (for acquiring and enjoying Leisure) are commonly also in the best interests of resident genes as 'fitness signals' in advertising mate quality to females (Miller 2000, 2009; Saad 2007). Attractiveness of a potential mate in this sense is typically interpreted to be correlated with his prospects (through genetic bequeathal) for resourcefulness (including through creativity) or for providing protection—thus addressing survival needs for oneself and one's offspring. And as a product of evolution, it is correlated then with his prospects for passing on these adaptive traits to male offspring (Chap. 7). An interesting (and unexplored) extension here is to ask whether these displays are attractive in part because they also signal a potential mate who is well-equipped in deploying delusions for 'extension of self' (Legacy needs) and distractions for 'escape from self' (Leisure needs), thus representing a good prospect as a positive, uplifting companion (also advertised, truthfully, through smiles and laughter), and for helping to raise offspring that are similarly well-equipped (through genetic bequeathal) with the Legacy and Leisure Drives needed to keep self-impermanence anxiety and other inevitable human disquietudes at bay. Of course, in none of the above does the adoring female need to be aware that her attraction to the potential mate has been informed by genetic inheritance, nor that its consequence is likely to affect her own gene transmission success.

Conspectus

The individual's most vital need is to prove his worth, and this usually means an insatiable hunger for action. For it is only the few who can acquire a sense of worth by developing and employing their capacities and talents. The majority prove their worth by keeping busy. A busy life is the nearest thing to a purposeful life. But whether the individual takes the path of self-realization or the easier one of self-justification by action he remains unbalanced and restless. For he has to prove his worth anew each day. It does not require the uncertainties of an outlandish doctrine of pre-destination to drive him to frantic effort and a striving to do something. (Hoffer 1976)

I have argued here (and in the previous two chapters) that the need for existential meaning, defined here in terms of 'self-impermanence anxiety buffers', has conceptually unique implications for reproductive fitness. Contrary to Schaller et al. (2010), the need for existential meaning does not merely '…operate in service to each of the other fundamental human needs'. I agree however, with their conclusion that '…these psychological desires … as with so many other human aspirations and goals, offer a means to facilitate our survival and reproductive fitness, and thus the reproduction of our genes in our children, and in our children's children, too' (Schaller et al. 2010). For a sentient species with a theory of mind, and an instinct for possessing a mental life that exists separately from material life—and one that can foresee the eventual annihilation of both, and feel a crippling anxiety because of it (Chap. 4)—Leisure Drive and Legacy Drive, I suggest, are critical for gene transmission success. They serve an intrinsic domain-general need: to be at least periodically distracted from this foresight, and/or at least periodically fooled into thinking that, despite knowledge of a mortal body, the inner 'self'—the memetic manifestations of one's mind—can transcend death. Meet *Homo absurdus* (Aarssen 2018, 2019).

When evolution gave our ancestors self-impermanence anxiety, it also gave them a mechanism for managing it: a drive for producing offspring as copies of 'self'—a delusional perception of post-self, symbolic immortality available through shaping their minds to mirror the minds of their parents. In an odd twist of irony then, fear of failed legacy became an adaptation (Aarssen 2010). Offspring became not just incidental products of sex (as with other animals)—and hence vehicles for genetic legacy—but also vehicles for memetic legacy. Importantly, motivation for the latter boosted the former, and (for our distant ancestors) *without there needing to be any awareness of genetic legacy*. This deeply ingrained drive for 'extension of self' also spilled over into other domains for memetic legacy involving accomplishment and religion—especially for males who were never really as certain of their paternity as they wanted to be (thus inspiring them to 'charge off in pursuit of fortune or fame or an imagined Utopia'—McGinley 1956); but also for young people anxiously waiting to have their first child, and for those hoping to have more children (or grandchildren) and wondering/worrying about why they didn't already. Religion and accomplishment thus eased their worried minds by providing supplemental domains for delusional memetic legacy, hence helping to ensure that their despondence (about uncertain paternity, childlessness, or having only one child or a small few number) did not cause them to give up trying. And grandparents were also there of course to spur them on: 'Children's children are the crown of old men' (King Solomon, Proverbs 17:6, King James Version).

The emotional cost of this fear of failed legacy (as with other aches and pains) is likely to be felt more acutely in the advancing years of older age—because of 'antagonistic pleiotropy' (Williams 1957). As Dobzhansky and Allen (1956) put it: 'The infirmities of old age are easily accounted for by the theory of natural selection. What happens to the organism after the reproductive age is of no concern to natural selection, or only in so far as the condition in old age is correlated with some traits which appear during the reproductive age'. In other words, feeling mortality and self-impermanence anxiety more sharply in later life necessarily imposes a decreasing penalty on evolutionary fitness, because normally by this time (at least for most of our senior citizen ancestors), gene copies had already been transmitted to the next generation. And so, while delusions of legacy (and distractions of leisure) may persist beyond middle age in defining psychological needs (e.g. Tinsley and Eldredge 1995; Iwasaki 2008), their effectiveness becomes much less (or un-) important for gene transmission success (especially for post-menopausal, hence infertile, women)—although these needs may serve to ramp up attraction to grandparenting (especially grandmothering for post-menopausal women), with attendant rewards for gene transmission success. Note also that only with recent advances in science and technology have humans become routinely capable of achieving average life expectancies as high as 80 years or more.

Importantly, as suggested above, when the 'wisdom of age' makes us not so easily deluded, Leisure Drive can still 'come to the rescue' (if one submits to it) by serving another intrinsic domain-general need: to be at least periodically distracted from the uniquely human agonizing uncertainty/suspicion (and for those so persuaded, from the conclusion of Darwinism) that we cannot transcend death, that there is no symbolic immortality or everlasting 'post-self', that legacy of the self is just a beautiful dream. Leisure Drive then, deployed in sufficient doses, protects us from learning, understanding, believing, and/or remembering that the only 'unifying purpose' or 'intelligent design' of life (if it can be called these) lies in the laws of physics and mathematics that shape the emergent properties of chemical, structural, and behavioural phenotypes. As Francis Crick (1994) put it: 'You, your joys and sorrows, your memories and your ambitions, your sense of personal identity and your free-will, are in fact no more than the behavior of a vast assembly of nerve-cells and their attendant molecules'. Humans, and apparently no other animals, have evolved not just the cognitive capacity to arrive at this diagnosis, but also a desperate need to purge it from consciousness.

The above tutelage from Crick then does not imply—as interpreted by philosopher Mary Midgley—that the self does not exist. It does not deny that humans are '… creatures with needs, tendencies and directions of their own' (Midgley 2014). It just says that the inner self, 'mental life', is not what our evolved psychology prompts us to hope that it can have: symbolic immortality. Both the hope for, and anxiety about failing to achieve this, nevertheless promoted genetic fitness of our ancestors. Importantly, evolution by natural selection tracks fitness, not happiness. As Nettle (2005) put it:

> The idea of happiness has done its job if it has kept us trying. In other words, evolution hasn't set us up for the attainment of happiness, merely its pursuit. ... We don't necessarily learn from experience that this is a trick, because we are not necessarily designed to do so.

In fact, as argued here, we are designed to be tricked, because this 'blind' pursuit, all by itself, has served well in propelling ancestral gene copies into future generations. Our motivations did not evolve to deliver us truth—only fitness. Consequently, much of human culture can be understood (and defined) as mostly just a collection of delusions and distractions that serve to reassure us that our existence is not absurd.

Components of Sexual/Familial Drive and Survival Drive, and their evolutionary roots (Box 9.1), are supported by a large body of literature (Chaps. 7 and 8). Legacy Drive and Leisure Drive, however, represent mostly hypotheses yet to be tested with more research. I predict that future studies will support their interpretation as essential for palliating the potentially incapacitating 'curse' of self-awareness—at least over the several years of reproductive immaturity required prior to successful mating and parenting. In this way, Legacy and Leisure Drives served to prevent the uniquely human fitness benefits of mental time travel and theory of mind (Chap. 4) from being compromised by self-impermanence anxiety. Recent advances in the field of 'terror management' theory—showing deployment of various mortality anxiety buffers, manifesting as behaviours that bolster self-esteem/meaning/purpose/redemption/value for one's life, and connected with a sense of membership within (and validation for) larger-than-self cultural worldviews (Greenberg et al. 2004; Vess et al. 2009; Pyszczynski et al. 2010; Solomon et al. 2010, 2015; Vail et al. 2010)—already point to the plausibility of the evolutionary interpretations argued here. As Solomon et al. (2004) put it:

> Thus, while cultures vary considerably, they share in common the same defensive psychological function: to provide meaning and value and in so doing bestow psychological equanimity in the face of death. All cultural worldviews are ultimately shared fictions, in the sense that none of them are likely to be literally true, and their existence is generally sustained by social consensus. When everyone around us believes the same thing, we can be quite confident of the veracity of our beliefs.

The four fundamental drives model developed here has potential for informing both theory and application for metrics of 'flourishing' and subjective wellbeing in positive psychology (Land et al. 2012; Wong 2012; Freire 2013; Leontiev 2013; Batthyany and Russo-Netzer 2014; Tay et al. 2014). Even more generally, it lays groundwork for a novel view of the evolutionary roots of human motivations and social life, and hence the rich and puzzling variety of cultural norms, celebrated across the globe, and underlining the scholarly interpretations of human history. In the next and in the final chapter, we will explore how a deeper and more broadly public understanding of these evolutionary roots will also be essential for achieving an effective response to the converging catastrophes of the twenty-first century.

> *The world of human aspiration is largely fictitious, and if we do not understand this, we understand nothing about man. … If you reveal the fictional nature of culture you deprive life of its heroic meaning because the only way one can function as a hero is within the symbolic fiction. If you strip away the fiction man is reduced to his basic physical existence—he becomes an animal like any other animal. And this is a regression that is no longer possible for him. The tragic bind that man is peculiarly in—the basic paradox of his existence—is that unlike other animals he has an awareness of himself as a unique individual on the one hand; and on the other hand he is the only animal in nature who knows he will die.* (Becker 1971)

References

Aarssen LW (2005) Why is fertility lower in wealthier countries? The role of relaxed fertility-selection. Popul Dev Rev 31:113–126

Aarssen LW (2007) Some bold evolutionary predictions for the future of mating in humans. Oikos 116:1768–1778

Aarssen LW (2010) Darwinism and meaning. Biol Theory 5:296–311

Aarssen L (2018) Meet *Homo absurdus*—the only creature that refuses to be what it is. Ideas Ecol Evol 11:90–95

Aarssen L (2019) Dealing with the absurdity of human existence in the face of converging catastrophes. The Conversation, 1 May, 2019, https://theconversation.com/dealing-with-the-absurdity-of-human-existence-in-the-face-of-converging-catastrophes-110261

Aarssen LW, Altman T (2006) Explaining below-replacement fertility and increasing childlessness in wealthy countries: legacy drive and the "transmission competition" hypothesis. Evol Psychol 4:290–302

Aarssen LW, Altman T (2012) Fertility preference inversely related to 'legacy drive' in women, but not men: interpreting the evolutionary roots, and future, of the 'childfree' culture. Open Behav Sci J 6:37–43

Angeles L (2010) Children and life satisfaction. J Happiness Stud 11:523–538

Batthyany A, Russo-Netzer P (eds) (2014) Meaning in positive and existential psychology. Springer, New York

Baumeister RF, Vohs KD, Aaker JL, Garbinsky EN (2013) Some key differences between a happy life and a meaningful life. J Posit Psychol 8:505–516

Becker E (1971) The birth and death of meaning: an interdisciplinary perspective on the problem of man, 2nd edn. The Free Press, New York

Becker DV, Kenrick DT (2014) Selfish goals serve more fundamental social and biological goals. Behav Brain Sci 37:137–138

Bellezza S, Paharia N, Keinan A (2017) Conspicuous consumption of time: when busyness and lack of leisure time become a status symbol. J Consum Res 44:118–138

Brooks D (2004) The new red-diaper babies. N Y Times. http://www.nytimes.com/2004/12/07/opinion/07brooks.html?ex=1260162000&en=ebdde83f03fe6d2e&ei=5090

Camus A ([1942] 1955) The myth of Sisyphus, and other essays (trans: O'Brien J). Knopf, New York

Caplan B (2012) Selfish reasons to have more kids: why being a great parent is less work and more fun than you think. Basic Books, New York

Choron J (1964) Modern man and mortality. The Macmillan Company, New York

Condon P, Makransky J (2020) Modern mindfulness meditation has lost its beating communal heart. Psyche, 16 Sept 2020, https://psyche.co/ideas/modern-mindfulness-meditation-has-lost-its-beating-communal-heart

Crick F (1994) The astonishing hypothesis: the scientific search for the soul. Touchstone, New York

Csikszentmihalyi M, LeFevre J (1989) Optimal experience in work and leisure. J Pers Soc Psychol 56:815–822

Davies C (2008) Single and happy: it's the freemales. The Guardian, 13 Apr 2008. https://www.theguardian.com/lifeandstyle/2008/apr/13/women.familyandrelationships3

Dobzhansky T (1961) Man and natural selection. Am Sci 49:285–299

Dobzhansky T, Allen G (1956) Does natural selection continue to operate in modern man? Am Anthropol 58:592–604

Freire T (2013) Positive leisure science: from subjective experience to social contexts. Springer, Dordrecht

Greenberg J, Koole SL, Pyszczynski T (eds) (2004) Handbook of experimental existential psychology. Guilford Press, New York

Hitz Z (2020) Lost in thought: the hidden pleasures of an intellectual life. Princeton University Press, Princeton

Hoffer E (1976) The ordeal of change. Buccaneer Books, Cutchogue

Iwasaki Y (2008) Pathways to meaning-making through leisure-like pursuits in global contexts. J Leis Res 40:231–249

Joyce K (2010) Quiverfull: inside the Christian patriarchy movement. Beacon Press, Boston

Julian K (2018) Why are young people having so little sex? Despite the easing of taboos and the rise of hookup apps, Americans are in the midst of a sex recession. The Atlantic, Dec 2018. https://www.theatlantic.com/magazine/archive/2018/12/the-sex-recession/573949/

Kaufmann E (2010) Shall the religious inherit the earth? Demography and politics in the twenty-first century. ProfileBooks, London

Kenrick DT (2011) Sex, murder, and the meaning of life: a psychologist investigates how evolution, cognition, and complexity are revolutionizing our view of human nature. Basic Books, New York

Kenrick DT, Griskevicius V (2013) The rational animal: how evolution made us smarter than we think. Basic Books, New York

Kenrick DT, Griskevicius V, Neuberg SL, Schaller M (2010) Renovating the pyramid of needs: contemporary extensions built upon ancient foundations. Perspect Psychol Sci 5:292–314

Land KC, Michalos AC, Sirgy MJ (eds) (2012) Handbook of social indicators and quality of life research. Springer, New York

Lenhart A, Smith A, Anderson M, Duggan M, Perrin A (2015) Teens, technology and friendships. Pew Research Center. http://www.pewinternet.org/2015/08/06/teens-technology-and-friendships/

Leontiev DA (2013) Positive psychology in search for meaning: an introduction. J Posit Psychol 8:457–458

Ma-Kellams C, Blascovich J (2012) Enjoying life in the face of death: east–west differences in responses to mortality salience. J Pers Soc Psychol 103:773–786

Maltby J (2010) An interest in fame: confirming the measurement and empirical conceptualization of fame interest. Br J Psychol 101:411–432

Martindale C (1980) Subselves: the internal representation of situational and personal dispositions. In: Wheeler L (ed) Review of personality and social psychology. Sage, Beverley Hills, pp 193–218

Maslow AH (1943) A theory of human motivation. Psychol Rev 50:370–396

McGinley P (1956) Women are wonderful: they like each other for all the sound, sturdy virtues that men do not have. Life Mag, 24 Dec 1956

Midgley M (2014) Are you an illusion? Acumen, Durham

Miller G (2000) The mating mind. Random House, New York

Miller G (2009) Spent: sex, evolution and consumer behaviour. Viking, New York

Nelson SK, Kushlev K, English T, Dunn EW, Lyubomirsky S (2013) In defense of parenthood: children are associated with more joy than misery. Psychol Sci 24:3–10

Nelson SK, Kushlev K, Lyubomirsky S (2014) The pains and pleasures of parenting: when, why, and how is parenthood associated with more or less well-being? Psychol Bull 140:846–895

Nettle D (2005) Happiness: the science behind your smile. Oxford University Press, Oxford

Pyszczynski T, Greenberg J, Koole S, Solomon S (2010) Experimental existential psychology: coping with the facts of life. In: Fiske S, Gilbert D, Lindzey G (eds) Handbook of social psychology. Wiley, London

Rowthorn R (2011) Religion, fertility and genes: a dual inheritance model. Proc R Soc B 278:2519–2527

Saad G (2007) The evolutionary bases of consumption. Lawrence Erlbaum, Mahwah

Schaller M, Neuberg SL, Griskevicius V, Kenrick DT (2010) Pyramid power: a reply to commentaries. Perspect Psychol Sci 5:335–337

Shaver PR, Mikulincer M (2012) An attachment perspective on coping with existential concerns. In: Shaver PR, Mikulincer M (eds) Meaning, mortality, and choice: the social psychology of existential concerns. American Psychological Association, Washington, DC, pp 291–307

Solomon S, Greenberg JL, Pyszcznski (2004) Lethal consumption: death denying materialism. In: Kasser T, Kanner AD (eds) Psychology and consumer culture: the struggle for a good life in a materialistic world. The American Psychological Association, Washington, DC

Solomon S, Greenberg J, Pyszczynski T, Cohen F, Ogilvie DM (2010) Teach these souls to fly: supernatural as human adaptation. In: Schaller M, Norenzayan A, Heine SJ, Yamagishi T, Kameda T (eds) Evolution, culture and the human mind. Psychology Press, New York, pp 99–118

Solomon S, Greenberg J, Pyszczynski T (2015) The worm at the core: on the role of death in life. Random House, New York

Taggart A (2017) If work dominated your every moment would life be worth living? Aeon, 20 Dec 2017, https://aeon.co/ideas/if-work-dominated-your-every-moment-would-life-be-worth-living?

Tay L, Chan D, Diener E (2014) The metrics of societal happiness. Soc Indic Res 117:577–600

Tinsley HEA, Eldredge BD (1995) Psychological benefits of leisure participation: a taxonomy based on their need-gratifying properties. J Couns Psychol 42:123–132

Trimberger EK (2005) The new single woman. Beacon Press, Boston

Trivers R (2011) The folly of fools: the logic of deceit and self-deception in human life. Basic Books, New York

Updike J (2000) The tried and the treowe. In: Due considerations: essays and criticism (2007). Random House, New York.

Vail KE, Rothschild ZK, Weise DR, Solomon S, Pyszczynski T, Greenberg J (2010) A terror management analysis of the psychological functions of religion. Personal Soc Psychol Rev 14:84–94

Vess M, Routeldge C, Landau MJ, Arndt J (2009) The dynamics of death and meaning: the effects of death-relevant cognitions and personal need for structure on perceptions of meaning in life. J Pers Soc Psychol 97:728–744

Williams G (1957) Pleiotropy, natural selection and the evolution of senescence. Evolution 11:398–411

Wong PTP (ed) (2012) The human quest for meaning: theories, research, and applications, 2nd edn. Routledge, London

Yaakobi E, Mikulincer M, Shaver PR (2014) Parenthood as a terror management mechanism: the moderating role of attachment orientations. Personal Soc Psychol Bull 40:762–774

Zeidler M (2018) Single women increasingly pursuing parenthood on their own. CBC News, 23 June 2018. https://www.cbc.ca/news/canada/british-columbia/single-women-increasingly-pursuing-parenthood-on-their-own-1.4717327

Chapter 12
Becoming the Solution

It seems to me true that Homo sapiens will never settle into any utopia so complacently as to relax all its conflicts and erase all its perversity-breeding neediness. It also seems true that the human species is heading toward a planetary triumph as complete as that of corn in Iowa; the paved-over and wired-up earth will produce a single crop, people, plus what people eat, and there will be nothing left of nonhuman nature for contrast or enlivening metaphor. (John Updike 2000)

John Updike (1932–2009) was one of America's most acclaimed writers and literary critics. It is easy for some to share his pessimism about the human condition and the probable fate of humanity, reflected in the above quotation. He may be right.

Some parts of this chapter are reproduced (with permission) from: Aarssen L. (2018) Meet Homo absurdus — the only creature that refuses to be what it is. Ideas in Ecology and Evolution. 11: 90–95.

But there are many more who are likely to be interested in finding reasons to be more optimistic, at least for the long term. The critical question then is: *Has our evolution equipped us to respond effectively to the converging catastrophes of the twenty-first century?* This is explored in this penultimate chapter, building on what we have learned in earlier chapters about the evolutionary roots of what we are. We are creatures with peculiar behaviours, social lives, and cultures, all informed by deeply ingrained motivations shaped in part by a long history of genetic inheritance, driven in each ancestral generation by the relentless but purposeless action of Darwinian selection. In the past, these motivational domains limited our capacity to learn from the history of our mistakes. They loomed large on the 'radar' of our ancestors, leaving little room for a deep understanding of the impact of humanity on our small planet, and essentially no room for significant motivation to protect it (Box 12.1).

Box 12.1: Depiction of the Four Fundamental Human Drives Occupying the Entirety of the Human 'Motivational Radar'

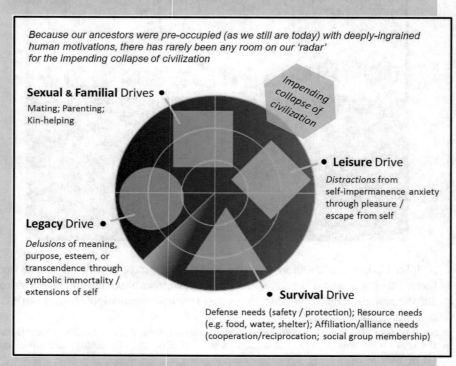

Because our ancestors were pre-occupied (as we still are today) with deeply-ingrained human motivations, there has rarely been any room on our 'radar' for the impending collapse of civilization

Sexual & Familial Drives •
Mating; Parenting;
Kin-helping

Impending collapse of civilization

• **Leisure** Drive
Distractions from
self-impermanence anxiety
through pleasure /
escape from self

Legacy Drive •
Delusions of meaning,
purpose, esteem, or
transcendence through
symbolic immortality /
extensions of self

• **Survival** Drive
Defense needs (safety / protection); Resource needs
(e.g. food, water, shelter); Affiliation/alliance needs
(cooperation/reciprocation; social group membership)

These motivational domains still loom large today. And our future, and the challenges that our descendants will face in it, will also be defined to a large extent by these evolutionary roots. But inaction can no longer be blamed on ignorance. From evolutionary biology, we now know what we are, and why. And if we care about the future prosperity of our species—or at least our immediate descendants—we have an obligation to carefully examine what we can expect to do about it, given what we are.

Our species has risen to many challenges in the past, but with consequences that usually generated still more, and even bigger challenges (Chap. 5). We have an impressive history of accomplishment in many ways, but many of its hallmarks—or their misappropriations—can only be judged today as regrettable (Diamond 2005; Dilworth 2010; Gluckman and Hanson 2019). Several, in recent decades, have written about how and why our planet is in peril, and what we need to do about it (Chap. 1). But most of these accounts have failed to recognize that the planet is not what is really most in need of being saved. The earth has recovered from catastrophes far worse than those unfolding now. It is *we* that are in need of being saved—from ourselves (Szent-Györgi 1970). This has always been the case—at least since the start of the agricultural revolution—but now that the earth is full, the task before us is more daunting than at any other time in the history of humanity. *We are the problem.* But importantly, by knowing this, and understanding why it is so, we have potential to become the solution.

A central mission of this book is to show how 'becoming the solution' involves three stages:

 (i) Understanding and facing what we are (Chaps. 6, 7, 8, 9, 10, and 11), and how we got that way (Chaps. 2, 3, and 4).
 (ii) Using this knowledge of the human condition to understand how it has shaped the history of human culture and civilization (Chap. 5), and why the latter is now on the verge of collapse (Chap. 1).
(iii) The hardest part: drawing heavily on our uniquely human capacities for intelligence, creativity, mental acuity, intentionality, and prosocial conscience to build a grand, revolutionary, global-scale cultural norm defined by a moral obligation to all of humanity, and hence to the biosphere and ecosystems of earth upon which human wellness depends. It sounds grandiose, and it is, but only because it is uncommonly large in scope; it is not absurd. And there is great potential value—with nothing to lose—in trying.

One thing cannot be ignored: some degree of collapse for civilization, as we know it, seems very probable in the coming decades (Diaz et al. 2019; Laybourn-Langton et al. 2019; Wallace-Wells 2019). How soon and how severe we cannot say with certainty. Many would say it has already begun. A 1973 production from the Australian Broadcasting Corporation (1973) predicted collapse by 2040. As Ronald Wright (2019) put it: 'Of one thing we can be sure: if we fail to act, nature will do so with the rough justice she has always served on those who are too many and who take too much'. With the above prescription, however, I believe it is possible, at least in theory, to ease the fall and facilitate the recovery—or at the very least, to design a new, more sustainable model of future civilization, one that has learned from our mistakes, and that will be ready for our descendants surviving in the aftermath of collapse. I believe we can achieve this if we learn how to better manage our genetic and cultural bequeathals—by understanding how to respond effectively to them. This will require doing a better job of embracing and cultivating some, while reigning in and saying 'no' to others. 'To arrive at where we want to be, we need to take full account of where we came from and of what we are' (Stamos 2008).

This is a tall order, and it will never even get off the ground without a broader public understanding of what our genetic bequeathals are, and why and how they need to be better managed. Unfortunately however—as world governments, economists, social scientists, and engineers grapple with an urgent global-scale crisis unlike any other in human history—the realities and implications of genetic bequeathals in human evolution are not even close to being on their radars. And it is no wonder why; Darwinian interpretations of human history and behaviour are still virtually absent from both elementary and secondary school curricula (Wiles 2006; Khazan 2019). Challenges also remain for post-secondary education (da Silva 2012; Wilber and Withers 2015). And we are fast running out of time. Meyer-Abich (1997) offers a call to action:

> Apparently, we need the industrial economy to treat problems we would not have without the industrial economy. More generally, we will need science and technology to treat problems that we would not have without science and technology. And we need to diffuse a new understanding of nature, including our own nature, in order to drive our science.

Revisiting the Crisis

Natural selection always maximizes gene transmission success (Chap. 2). This means that for any successful species, available resources are usually used to the fullest extent possible, and so birth rate generally rises as high as can be sustained. For wild species, this of course routinely increases crowding and so death rate is usually also relatively high, and this is commonly exacerbated further by the impact of pathogens and predators feeding on the abundance of prey. As long as birth and death rates are approximately balanced, population size stays relatively constant, at a level defined by the carrying capacity of the environment (i.e. the resource–supplying power available from the local habitat). This was the general state of affairs for *Homo sapiens*—with relatively high birth and death rates (Box 12.2, top panel)—sustained throughout practically all of the past 300 thousand years or so since our species' origin.

But emerging from the industrial revolution, things started to change dramatically (Chap. 5), and none of it would have happened without oil. (http://blogs.scientificamerican.com/plugged-in/2011/12/21/300-years-of-fossil-fuels-history-in-5-minutes/). Technologies, especially for agriculture and medicine, had advanced to a point where many more people (especially children) could avoid malnutrition and starvation, stay healthier, and recover more effectively from injury, illness, and infectious disease. Consequently, death rate started to decline, while birth rate (average number of births per woman per lifetime) stayed relatively high (Box 12.2, top panel). People not only lived longer, thus adding to population density, but more of them also survived long enough to reach reproductive age, and so there were more people having babies. Energy from fossil fuels, especially oil, and new technologies fueled by oil, meant that more mouths could be fed in each generation, and that still more therefore needed to be fed in the next. Population growth sky-rocketed, rising ever faster as the Anthropocene unfolded, with more and more oil pumping and flowing to power the escalating advances in agriculture, medicine, and technologies for urbanization.

Box 12.2: Where Are We Headed?

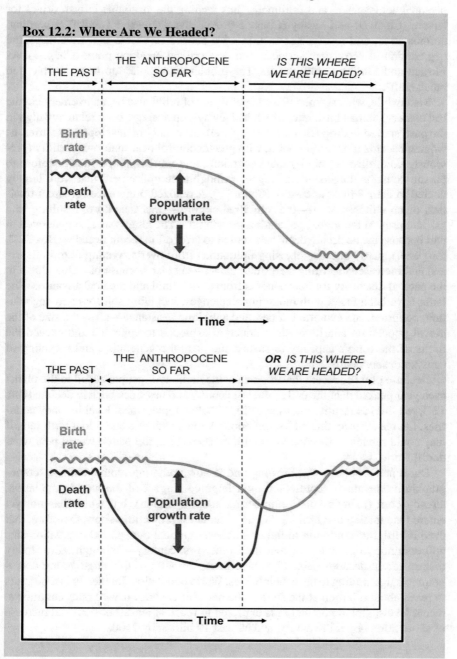

By the middle of the last century, many other new technologies for science and engineering (also enabled by oil) came online to kick-start the 'great acceleration'. This allowed the lives of people, especially in developed countries, to be filled with more comfort, more consumerism, more entertainment, and more opportunities for personal achievement and fulfilment, thus feeding the insatiable human drives for leisure (Chap. 9) and legacy (Chap. 10). And the aggressive fossil fuel burning, deforestation, and resource extraction rates needed to enable this have taken the degradation of ecosystem services to a now critical breaking point (Chap. 1). As Ehrlich and Ehrlich (2008) put it: 'The problem is simple: too many people, too much stuff'.

Meanwhile, what seemed like a hopeful sign of relief also began to emerge in the last century: human birth rate, which had always, on average, been relatively high in the past, started to drop (Box 12.2, top panel), especially in developed countries, as women became more empowered, with greater control over their own fertility (previously controlled largely by male coercion), and more opportunities therefore to pursue domains for leisure and legacy through accomplishment that were largely denied to their female ancestors (Chap. 10). A realized preference for small families, or no children at all—the 'childfree culture'—was thus born, resulting in a gradual drop in the global population growth rate. The latter was also generated in part because the declining death rate started to level off in recent decades (Box 12.2, top panel), partly because of rising limitations from overharvesting (e.g. in fisheries) and increased crowding on a small planet—but also because of a slow-down in the pace of discovery for new benefits from agricultural and medical advances. The latter have been faced with mounting challenges, including soil loss and degradation, pollution, and evolution of pest and antibiotic resistance—all as the size of the global population and its waste production exploded to approach and exceed the limits of the earth's capacity to sustain the natural regeneration and recycling of ecosystem services.

Because of these recent developments, the number of people added to the planet each year peaked near the end of the last century and has since been in decline (Box 12.3, red line) as fertility rates have dipped below replacement level in many countries. On the surface, this all looked good: A low death rate and a low birth rate, if they could remain in balance for the future (Box 12.2, top panel) would be a wonderful world.

But a perfect storm was brewing and is now gathering momentum—a 'demographic momentum', represented by a large contingent of the world population, already born (today's youth, numbering about 2 billion), but which has not yet entered the mating and birthing 'arena'. The latest projections show, therefore, that even if birth rate continues to fall at the recent gradual rate, global population size will continue to grow throughout the twenty-first century—although more slowly than in recent decades (Box 12.3, blue line)—because of this large young demographic bulge waiting to have their babies. World population doubled in just 40 years to reach about 6 billion at the start of the new millennium. If vital rates continue at recent levels, global population is expected to reach at least 9 billion—an increase of about 50%—by mid-century, and at least 10 billion by 2100.

Box 12.3: Despite a Declining Growth Rate, an Increase in Global Population Size Is Expected to Continue Throughout the 21st Century

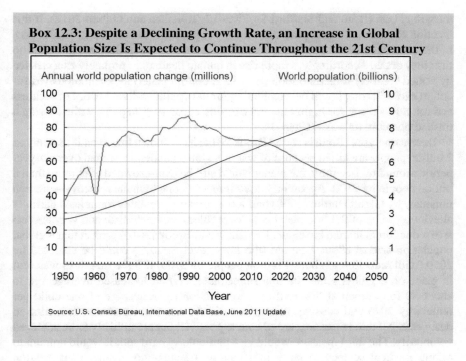

Source: U.S. Census Bureau, International Data Base, June 2011 Update

As our small planet approaches nine or ten billion people in this century, combined with ecosystem services degraded even more than they already are, it will not be a pleasant home for most residents. When the only alternative is an endless escalation of local misery, even good, law abiding citizens, when pushed to the limit, inevitably choose, eventually, to raid neighbours who have more resources. Some think that all we need to do is ramp up our space program so that we can expand our business-as-usual 'March of Progress' to another planet. But it's not going to happen, at least not in time,[1] and probably never.

The level of misery in the coming decades, therefore, will depend on the success of several strategies and remedies that we already know have potential for slowing down the rate of collapse or minimizing the impact of its effects (e.g., Homer-Dixon 2007; Speth 2008; Heinberg 2011; Barnosky et al. 2012, https://steadystate.org/). This will depend critically on our capacity to vigorously reduce per-capita consumption rates, and to correct and reverse, as much as possible, the existing degradation of ecosystem services. Unfortunately, as we saw in Chap. 1, public and political will for this has never been significant on a global scale and remains today

---

[1] Miller (2007) offers the sombre speculation that the reason why we have not yet made contact with other advanced extraterrestrial life is not because we are alone in the galaxy, but because evolved technical intelligence (like ours) has some deep tendency to be self-limiting, even self-exterminating.

meagre at best (Brito and Stafford-Smith 2012; Tollefson and Gilbert 2012). If this inaction continues, the root problem—too many people—will correct itself abruptly, in the same way as for populations of all wild species when they get too big, too fast: they crash. Accordingly, a rapid rise in human death rate (probably exacerbated by armed conflict) sometime soon (Box 12.2, bottom panel) looks ominously diffi-cult to avoid. If true, it will be a reckoning for our centuries-long history of reckless and relentless reach for higher human carrying capacity (Chap. 5), while leaving a trashed planet in its wake (Chap. 1).

Recent analyses (Bradshaw and Brook 2014) have shown that even if as many as 2 billion people are killed because of war or a global pandemic, spread over a 5-year period around mid-century, the planet would still be over-crowded with more than 8 billion people by 2100. Apparently, therefore, only one humane remedy—foreseen famously by Garret Hardin (1968) half a century ago—remains for at least partially alleviating some of the 'too many people' problem: an immediate and very aggres-sive reduction in human birth rate. Bradshaw and Brook (2014) project that a global population size of about 7 billion (the same level as 2013) could be achieved by 2050 (with reduction to about 4 billion people by 2100) if three measures were put in place on a global scale, starting immediately: (i) an increased average age to about 30 for women at first birth; (ii) a reduction to an average of one child per female by 2045 and constant thereafter to 2100; and (iii) no further reduction in death rate. All three measures are of course possible, but are unlikely to be achieved voluntarily. The third measure would be a particularly hard sell; it would amount to asking medical science to lapse into merely a status-quo management, to stop researching how to increase cures, recoveries, and survival rates from disease, and to abandon research that promotes slower aging and life span extension. This would be a piercing affront to our intrinsic human drives for both Survival and Legacy. Most will vote for politicians that promise easy, future availability of anti-aging drugs/therapies and life-span extension procedures, especially if this includes potential for good health and vitality (Bloom et al. 2015). Recent research has already started to deliver on these prospects for consumers (e.g. Hayden 2015).

Who Will Be the Parents of the Future?

Going forward, a critical question looms: Will human birth rate continue to drop *voluntarily* and substantially in the coming decades—both in developed countries, where more and more women are embracing small families and the 'childfree' cul-ture (see Chap. 10), and in less developed countries, where empowerment for women is gaining more and more traction (Aarssen 2005; Engelman 2010; Vollset et al. 2020)? We should hope so, but this may be unlikely to unfold so neatly, and maybe not at all. As discussed in Chap. 11, the 'childfree' culture may be just a short-term rebound effect of recent female empowerment, where attraction to off-spring production is commonly low because it was also true of many female ancestors—but who were nevertheless forced (because of subjugation by males) to transmit their genes (that informed a weak maternal instinct) to future generations.

One thing is certain: the motivations of those in future generations will include the same motivations that reward reproductive success in this generation. In other words, parents of the future will not be products of the childfree culture, nor will they be products of the 'freemale' culture, or the 'sex-recession' culture (see Chap. 11). Most of them will be the descendants of women today who are anxious (or at least content) to raise children, and probably especially those who are anxious to raise a lot of them, and so to be very interested in sex with partners who are interested in helping to raise them—because of a strong 'parenting drive' (Chap. 11), and/or other traits that may be correlated with high fertility, informed in part by genetic inheritance (e.g. 'dumbing down' effects; Nuwer 2012; Kanazawa 2014; see the first 10 minutes of the film 'Idiocracy' (2006)). Accordingly, selection against weak 'parenting drive' (which had little opportunity to operate under centuries of patriarchal subjugation of women) may soon be ramping up now that more and more women worldwide are becoming free to decide for themselves whether they will have children, and how many—thus reversing the current de-population trend resulting from below-replacement fertility, and so making a stable future population size more difficult to attain than what some projections have suggested. A recent UN report (UN, Department of Economic and Social Affairs, Population Division 2019) indicates that '… even in populations with low or very low fertility for decades, women continue to express a desire for around two children on average", and that "… a rebound in the fertility trend in current low-fertility countries seems the most plausible future trajectory over the long run'.

Natural selection never limits the reproductive success of resident individuals of any species in order to minimize the collective misery of over-population. Echoing Charles Galton Darwin (1953), Dobzhansky (1962) warned, over half a century ago:

> *Reduction of the birth rates is necessary if the population growth is to be contained. Family planning and limitation are not, however, likely to be undertaken by everybody simultaneously. Those who practice such controls will contribute to the following generations fewer genes in proportion to their number than those who do not. Fewer and fewer people will, therefore, be inclined to limit their families as the generations roll by. The human flood, rising higher and higher, will overwhelm a multitudinous but degenerate mankind. The assumption implicit in this argument is, of course, that the craving for perpetuation of one's seed is uncontrollable by reason and education, and that people will go on spawning progeny, even knowing that it is destined to be increasingly miserable.*

Accordingly, those with the highest genetic fitness are always those that contribute the most to causing overpopulation. In other words, the benefit of producing an additional reproducing offspring (higher genetic fitness) is always realized exclusively by the parents. But if the cost of this extra offspring production (increased crowding) is shared with other parents, then the per capita magnitude of [benefit-cost] is always greater for individuals that produce more than one reproductive offspring than for other individuals in the same population that produce only one. Natural selection, therefore ('left to its own devices'), will always favour those individual traits that promote high fecundity relative to neighbours, even among impoverished individuals that are already crowded at carrying capacity. Overpopulation, then, is an inevitable product of natural selection—as in Hardin's (1968) 'tragedy of the commons': 'Commonizing' costs (overpopulation) while 'privatizing' profits (genetic fitness).

It seems inevitable (at least reasonable), therefore, to speculate that without cultural selection for mutually agreed coercive measures to limit the number of births per female, contemporary pro-natalist cultures (e.g. like 'Quiverfull'; Joyce 2010), and cultural products of 'dumbing down' (Nuwer 2012; Kanazawa 2014), could soon displace the 'childfree' culture, resulting in rising birth rates in the coming decades—especially in response to a rise in death rate and population crash, as presently forecasted.

We are now approaching the end of oil, and with this, it is hard not to envisage severe limits for future advances in agricultural and medical technology. Can we count on things like genetically modified crops (Nature 2013), and vertical (highrise) farming (The Economist 2010) to grow and provide more and more food to feed more and more people? Can we count on medical science to stay ahead of the now alarming and growing wave of antibiotic resistance (Spellberg 2009; McKenna 2013; World Health Organization 2014), and pandemics (like Covid-19)—all without oil? If technology fails to rescue us from this—*and ironically, especially if it does*—will both the death and birth rates of humans not return soon, in balance, to the high levels similar to what most of our ancestors endured since the origin of our species (Box 12.2, bottom panel)? Thomas Malthus (1798) saw it coming over two centuries earlier. The crisis took much longer to unravel than what he anticipated, but only because the power of oil was unknown at the time, and so he could not foresee the technological revolutions that were about to unfold because of it. Malthus' prophesy, however, is now upon us as the oil runs out (Lidicker 2020), leaving us also with catastrophic global warming effects (Tollefson 2018; Ripple et al. 2020; UN Environment Programme 2020) from over a century of pumping it out and burning it up as fast as we could. We need a plan.

Onward Biocultural Evolution

Genetic evolution and cultural evolution have both failed to produce human motivations and inclinations capable of fostering a sustainable global civilization with limitations on population size, moderate consumption, and preservation of ecosystem services (de Duve 2010). Is it reasonable then to expect that either can be called upon now to come to our rescue in the twenty-first century? Dobzhansky (1973) ponders:

> The past cannot be changed regardless of our judgment, but man is no longer obliged to accept future evolution caused by blind and impersonal forces of nature. Evolution may eventually be managed and directed. Must it go on in the same direction in which it went in the past? Possibly so, yet only provided that this direction appears, in the light of human wisdom, good and desirable. It is not good by definition.

Clearly our best hope—especially because time is of the essence—lies heavily in culture (Ehrlich and Ehrlich 2013). Still, genes can 'hold culture on a leash' (Wilson 1978). Many of the human motivations and behaviours that cause us trouble today are the same as those of our predecessors that were leaving the most descendants (de Duve 2010). Some of their genes that we thus inherited are likely, therefore, to be 'whispering' harder than others, and so the motivations and behaviours that they inform may not be so easily manipulated by social learning. But as we saw in Chap. 6, the relationship between genes and culture normally functions as a fluid inter-dependence, including with culture holding genes on a leash. The latter is particu-larly conspicuous when social learning (e.g. laws against crime) compels us to say 'no' to inclinations informed by 'whispering genes' (e.g. temptation, from greed, to steal). As Barash (2008) put it: 'Human beings are capable not only of understand-ing the route along which evolution has placed them, but also of deciding, in the clear light of reason as well as ethics, whether to follow this path.'

Importantly, therefore—contrary to popular opinion within some circles—Darwinism does *not* imply that we cannot be held accountable for our actions. A promising plan for the future must entail a deep understanding of the dynamics and potential of biocultural evolution (Chap. 6). Kingsolver (1995) proclaimed a rally-ing call:

> We humans have to grant the presence of some past adaptations, even in their unforgivable extremes, if only to admit they are rocks in the stream we're obliged to navigate. … A thou-sand anachronisms dance down the strands of our DNA from a hidebound tribal past, guid-ing us toward the glories of survival, and some vainglories as well. If we resent being bound by these ropes, the best hope is to seize them out like snakes, by the throat, look them in the eye and own up to their venom.

How then might we expect social learning to come to our rescue? Technology would of course be part of it, but it would need to have a very different agenda from the one that drove the 'March of Progress' for civilization (Chap. 5)—i.e. feeding more and more people—over the past 10,000 years or so. If the 'problem is sim-ple'—'too many people, too much stuff' (Ehrlich and Ehrlich 2008)— then the solution is equally simple (at least conceptually): less people, less stuff (https://www.earthovershoot.org/. Many of the details for this solution are already well-established, involving two main agendas for sustainability in the twenty-first cen-tury: (i) a 'Population Ecology Revolution', where solutions lie not in feeding more and more people, but in limiting the number that need feeding (Cafaro and Crist 2012; Butler et al. 2015; https://populationspeakout.org/, https://populationinsti-tutecanada.ca/); and (ii) an 'Agroecology Revolution', where solutions lie not in feeding the world, but helping and allowing the world to feed itself with locally sustainable, ecologically sound systems combined with equitable food distribution (Martin and Sauerborn 2013).

The 'Green Revolution' of the last century and the 'Biotechnology Revolution' currently in vogue are based on a model of industrial agriculture, which aims to alter natural ecosystems to maximize short-term yields of a few specialized monoculture crops for export to world markets, while maximizing profits to big-business at the expense of lost biodiversity and destruction of ecosystem services. In contrast, the

'Population Ecology Revolution' combined with the 'Agroecology Revolution' is based on achieving a stabilized population size, and achieving sustainable crop yields, while preserving biodiversity, protecting ecosystem services, and emphasizing locally produced food for local people with local control. It is an ecological approach that views humans as products of, and partners in, the local ecology, not masters of it—and as partners in a 'plenitude economy' (Schor 2010) that fosters moderation in per-capita consumption. In short, it is a viable model for anticipating a future that could, someday, look like the top panel in Box 12.2. Under this scheme, a sustainable human carrying capacity of the earth, with an acceptable human development index for all, and while preserving natural ecosystem services, will be only a fraction of the present global population size—probably on the order of about two to three billion people according to some estimates (Pimentel et al. 2010). Importantly, it's not just about climate change. It's not even mainly about climate change. As Paul Ehrlich (1968) put it: 'Whatever your cause, it will be a lost cause without population control'. From Burr (2008):

> *Exponential population growth will overwhelm all efforts to create a sustainable society. Improve auto fuel efficiency by half but double the number of cars and you are back to where you started. Increase food production and more people will eat it up. Preserve open space here, and development will still occur some place else. Cut greenhouse gas emissions in half, but double the population and all of the gains are wiped out.*

The science and engineering needed for these new revolutions are ready to do their work (Nekola et al. 2013). All that is needed is to prepare the way for cultural selection to do *its* work, through carefully guided social learning, peaceful activism, and reform. Henry Mitzberg (2014) has highlighted some essential components for this:

> *Few governments today take the lead in addressing our serious problems, such as poverty amidst plenty and the degradation of our physical and social environments. Most have been either co-opted or overwhelmed by the implicit alliance that has formed of corporate entitlements with the economic dogma. Moreover, most of our governing institutions are stuck in an 18th century view of democracy that caters to individual demands, whereas many of our problems require collaborative efforts to address common needs, globally as well as domestically. ... The radical renewal we require will need to rely, especially in its early stages, on community groups and associations of the plural sector. ... whereby social movements and activist associations challenge destructive practices that we can no longer tolerate. Such movements are on the rise, and can accomplish a great deal—if they can focus their efforts, with sharper tactics.*

Success for these efforts, however, will ultimately require that the social sciences participate in a critical public service: *educating the public about the evolutionary roots of what we are, and from this, how to manage our genetic bequeathals so that some of them do not impede—and so that some of them can serve—the capacity for biocultural evolution to come to our rescue.* Parallel movements within the arts and humanities will also be needed to contribute to this mission. The social learning involved here will need to have impact on everything from new school curricula, political agendas, and government policies, to revised narratives of history, pop-cultural manifestos, journalism and social media, story lines and themes of literature and film, and inspirations of music and other products of art. Alexander (1971) ruminates: 'I suggest that the only effective way for man to discard any part of his evolutionary background that he decides he does not like is by first understanding it thoroughly —by lifting it into his conscious consideration'. If we ignore this, it will leave our evolutionary roots and contemporary forces of natural selection to have their way with us in shaping more of the same catastrophic 'business as usual' for civilization in the twenty-first century.

Importantly, there is no reason to believe that biological evolution has ever slowed down or that it is slowing down now (Stearns et al. 2010; Bolund et al. 2015). As Dobzhansky and Allen (1956) put it: 'Natural selection would cease only if all human genotypes produced numbers of surviving children in exact proportion to the frequencies of these genotypes in the population. This does not, and never did, occur in recorded history'. Recent accounts in fact indicate that human evolution has probably accelerated because of the advance of civilization, particularly over the past 10,000 years (Cochran and Harpending 2009). But natural selection is without purpose; it can never deliver to us a peaceful, sustainable civilization—only differential gene transmission success. Our motivations and other cognitive capacities did not evolve to give us a promising future—only fitness. These are our deepest evolutionary roots. Their hold on us has been modulated by cultural evolution over at least the past 40 thousand years or so, but not in ways that have equipped us to respond effectively to the converging catastrophes of the twenty-first century. Cultural selection can now come to our rescue, but only if given a new and powerful agenda for its goals.

Biosocial Management Goals

Human wellness requires provisions from healthy ecosystem services, but more generally it requires provisions that address the core intrinsic human needs and motivations that have been shaped in part by genetic inheritance. We are stuck with these genetic bequeathals, at least for now. And so if biocultural evolution is to come to our rescue in time, it will require what might be called 'biosocial management'. Starting points can be found in the new and expanding field of 'applied evolutionary psychology' (Saad 2011; Roberts 2012).

Biosocial management will involve a two-pronged approach: (i) ensuring that there are effective domains in place within cultural norms and institutions to sufficiently appease or mitigate the 'big four' human drives (Box 9.1); but at the same time, (ii) ensuring that these provisions do not harm other people or the planet; i.e. they do not compromise the principles for achieving a sustainable, environment-friendly, prosocial model for civilization (culture 'holding genes on a leash'). The central tenet then of biosocial management is this: *neglecting the former will compromise success for the latter—because genes can also 'hold culture on a leash'*. It amounts to making room on our 'radar' for both components (Box 12.1).

For some of our genetic bequeathals, effective biosocial management will be particularly challenging:

Addiction to Consumerism

Conspicuous and wasteful consumerism feeds several intrinsic human needs. It can signal membership within a particular in-group, addressing the need for self-esteem or reputation legacy, and it can advertise status that is attractive to potential mates. Nettle (2005) illustrates the conundrum in this:

> We are satisfied with our income or the size of our car in comparison with the incomes and cars we see around us. ... We have evolved in environments where there were numerous possible ways of making one's living, and our reproductive success would have been dependent not on absolute wealth but on relative status. Since it was impossible to know inherently what the optimal behaviour in local conditions would be, we evolved a psychology of looking at those around us who seemed to be doing best, and trying to do even better than them.

Wasteful consumerism can also play out in the culture of 'cool' (Chap. 10), and especially in driving the fashion industry, and high culture. Saad (2007) explains:

> From an evolutionary perspective, the dynamics driving the fashion industry and the corresponding snob and bandwagon effects are similar in spirit to those between pathogen and hosts locked in a classic evolutionary arms race. As soon as the fashion laggards have 'inoculated' themselves against the new virus (i.e. by adopting the latest fashion trend), the innovators have 'mutated' onto a new fashion trend. At a more macro level, this particular set of dynamics can easily be used to explain the adoption of new symbols of high culture. As objects of elite culture diffuse onto the masses and hence no longer serve as accurate indicators of one's social standing, new symbols must be found that delimit membership into the various strata. Art movements might be construed as an instantiation of the latter phenomenon. Art connoisseurs create a new trend ... which sets in motion the evolutionary arms race meant to clearly delimit the in-group and out-group members.

The 'rat race' of capitalist materialism and consumerism— where the average person never seems to really get ahead or to find greater happiness or contentment, despite continuing advances in technology, economic growth, and personal prosperity (Easterlin et al. 2010)—has an intriguing parallel in the 'Red Queen hypothesis' of natural selection. According to the Red Queen, from *Alice in Wonderland* (Lewis Carroll 1872), it takes 'all the running you can do just to keep in the same place'— and on average, an individual will leave only one descendant, despite that natural selection continually favours individuals that leave more Harper (1977). Similarly, it seems, in the current project of human civilization, it takes all the materialism and consumerism that you can get just to stay content (including with sufficient distraction from self-impermanence anxiety). Within developed countries, this has humanity locked into a kind of 'hedonic treadmill' (Brickman and Campbell 1971), where we seem driven to stay committed to endless economic growth and the resource extraction and energy consumption needed to support it. As Tennessee Williams (1955) wrote: 'The human animal is a beast that dies and if he's got money he buys and buys and buys and I think the reason he buys everything he can buy is that in the back of his mind he has the crazy hope that one of his purchases will be life ever-lasting'.

On Being Male

Virtually everything that humans have screwed up on this planet has been delivered at the hands of men. Most proximately this is because men simply denied opportunity for women—essentially because they could; i.e. men generally could physically overpower women more than the other way around. But more fundamentally, men were generally *inclined* to control the activity of women more than the other way around. And this is because of the evolutionary consequences of this control: our male predecessors who achieved this more forcefully and on larger scales generally left more descendants (Chap. 7). Hence, these male motivations were fueled not just by social learning and cultural transmission, but also in part by 'whispering genes' that informed a routinely worrying paternity uncertainty, and a commonly hyper-active male sexual impulse, characterized by perpetual vulnerability to sexually arousing distractions and sexually charged bravado that has always been, on average, much more acute than in females.

These male dispositions account for many of the hardships, costly mistakes, and atrocities in human history. This includes everything from the long history of subjugation and abuse of women to the male cultures of costly fitness signalling, aggressive competition for access to mates, including violent defence of social reputation, and motivation for going to war (Wrangham and Peterson 1996; Bribiescas 2021). This dark legacy, still with us today, not only continues to be a monumental abomination against women, but it also imposes a major limitation on the capacity of societies to respond effectively to the converging catastrophes of the twenty-first century. Future success for biosocial management, and in particular the capacity for men to play an effective role in directing it, will in large measure depend on finding and implementing effective, socially responsible, and gender-respectful remedies for mitigating this vexing disparity in the sexual inclinations, behaviours, and cultures of men and women (see also Baumeister 2010; Bribiescas 2011).

Us Versus Them

> The need is not really for more brains, the need is now for a gentler, a more tolerant people than those who won for us against the ice, the tiger and the bear. The hand that hefted the ax, out of some old blind allegiance to the past, fondles the machine gun as lovingly. It is a habit man will have to break to survive, but the roots go very deep. (Loren Eiseley 1957)

Throughout recorded history to the present day, the proclivity to go to war (Chap. 8) has largely persisted (Braumoeller 2013), and some analyses suggest that the frequency of military conflicts has increased since the nineteenth century (Harrison and Wolf 2012). This of course presents a serious impediment to biosocial management objectives. Perhaps it is the most concerning of all—because the global environmental crisis itself (especially climate warming) and associated resource shortages (especially for oil and water) can be expected to drive and intensify future conflict between nations (Dyer 2008). It already has (Morello 2013). Responding to the collapse of civilization is challenging enough by itself. But as an escalating reason for going to war, effective response to collapse is increasingly encumbered by the rising costs of escalating engagement in, and consequences of, war. Additional reasons also make it difficult to be optimistic about ever being able to eradicate war. Smith (2007) expounds:

> The act of killing is supremely rich in sensory input. Think of a lion bringing down a zebra. There is the taste of blood, and the smell of blood, the prey's helplessness, the struggle, and its shrill cry of distress. These sensations must have excited prehistoric humans, too, and with sexual rewards in the offing, we can add mounting erotic tensions to the mix. War is simultaneously hideous and exciting. Once in the grip of the illusions of war, men take pleasure in killing. ... It is horrible because of the danger and suffering that soldiers and civilians endure, and the unavoidable guilt that comes from killing. It is pleasurable because—like all pleasure—it is something that benefitted our ancient ancestors who were victors in the bloody struggle for resources. The joy of war is the joy of the hunt, of bringing down game, of ridding the world of a man-eating monster or obliterating a plague. ... Our relationship with killing is ambivalent, a compound of pleasure and aversion. Both are deeply rooted in human nature, and neither can be extirpated. If I am right, we will never stop men from enjoying war, and trying to do so is a fool's errand.

War has always provided an effective domain for Legacy Drive in men (Chap. 10)—a vehicle for purpose or 'meaning' that almost welcomes death—because the great comradery of war delivers a vital and vibrant sense of belonging to a purpose 'larger than self'. Choron (1964) explains:

> Life dedicated to revolutionary combat acquires another dimension in which ... isolation and loneliness is replaced by the sense of belonging, of brotherhood with other men, and death in the fight for a common cause becomes meaningful.

An enormous challenge, perhaps insurmountable, for biosocial management (and social and political management in general) lies in building new and more non-violent cultures that espouse a sense of larger-than-self identity and group membership that make war look unnecessary.

Self-Deception

Successful biosocial management will require careful, precise, and honest thinking. Evolution has given humans cognitive capacities of many kinds that will help to facilitate this, but this bequeathal has not included a filter for self-deception. As Gustav Le Bon (1896) famously said:

The masses have never thirsted after truth. They turn aside from evidence that is not to their taste, preferring to deify error, if error seduce[s] them. Whoever can supply them with illusions is easily their master; whoever attempts to destroy their illusions is always their victim.

Self-deception (cognitive bias) makes it easy to be complacent, distracted, deluded, or in denial about unpleasant things, like inevitable mortality, climate change, or an impeding collapse of civilization. The dilemma for biosocial management then is that we cannot afford to be complacent or in denial about the latter, but at the same time we have evolved fundamental needs for delusions of Legacy (Chap. 10) and distractions of Leisure (Chap. 9) in order to manage self-impermanence anxiety. In fact, adaptive cognitive biases are evident across many domains (Haselton et al. 2016), including inferences about risks to survival, (Chap. 8), the trustworthiness of others (Chap. 8), and the sexual and romantic interests of prospective mates (Chap. 7). In other words, cognitive biases are '… not necessarily *design flaws*— instead, they could be *design features*' (Haselton et al. 2016).

Unfortunately, however, people commonly discount valuable expertise when it is associated with any opinion that they would rather not hear (Kahan et al. 2010). Ironically, in one sense, this makes science inherently heretical; in other words, science seeks to see the world as it is, not as we would like it to be. The latter of course is commonly delivered nevertheless through the comforting myths of culture, which, as Diamond (1992) illustrates, have commonly been responsible for several hundred years of delay in the public acceptance of science: e.g. concerning heliocentrism (i.e. the earth is not the centre of the universe), concerning germ theory (i.e. human diseases and epidemics are not divine judgements on the wickedness of humans), and concerning evolution (i.e. humans were not specially created by God, and the earth was not made for humans and does not belong to humans). In many respects, '… cultural evolution is just one damned delusion after another' (Aarssen 2010).

Our current cultural myth is to deny that both population and economic growth are the causes of environmental degradation, because you need economic growth in order to feed the addiction to consumerism and materialism, and you need population growth in order to have economic growth (New Scientist 2008; Washington 2020). As Burr (2008) put it:

In the twilight of the age of exuberance the idea that there might be limits to growth is for many people still impossible to imagine. Limits are politically unmentionable and economically unthinkable. Our culture tends to deny the possibility of limits by placing a profound faith in the powers of technology, the workings of a free market, and the growth of the economy as the solution to all problems, even the problems created by growth.

This is part of a bigger, very stubborn myth that human history has been a long tale of benefits accrued from progress and growth. And we appear to be a long way from achieving public awareness and acceptance of science from anthropology and archaeology which shows, in contrast, that human history—for most of our ancestors, and still for most people alive today (as for all species)—has been a long tale of impoverishment and misery from resource shortage and starvation, and for humans in particular, also from warfare and genocide.

Self-deception then commonly involves equipping the mind with optimism. And to be a good optimist one needs to be a good liar. All of it routinely rewarded ancestral gene transmission success. Smith (2007) explains:

> A liar who believes his own lies is far more convincing than one who doesn't, so if you sincerely believe that you are telling the truth, there is no reason to be awkward and self-conscious. The accomplished self-deceiver can deliver self-serving falsehoods without even breaking a sweat. … In the social world, where deception and manipulation often rule, it sometimes pays not to know too much about your own agenda. In these circumstances, the person who is too insightful often ends up the loser.

Researchers are intrinsically drawn to progress-based optimism, which is where some of the greatest opportunities lie for satisfying our intrinsic (yet delusional) attraction to legacy (Chap. 10). As Hardin (1999) muses, researchers are not drawn to discoveries that '… undermine an important secular element of faith in our time, namely that important discoveries in science always improve the human situation in the natural world. … Improvement is an optimistic idea; research students expect to benefit financially from optimistic discoveries. Disimprovement is pessimistic; canny students doubt that they will benefit from being part of a team that makes pessimistic discoveries. Legions struggle to get through the door of optimism; the doorway of pessimism is uncrowded'.

It will be important, therefore, that research and applications in biosocial management be understood in terms of optimistic improvement, even though it involves confronting a collapsing civilization and hence the realization that discoveries of science have not always improved 'the human situation in the natural world'—and have in fact failed miserably to equip humanity with prosperity for the twenty-first century.

At the same time, however, we are challenged by the irony that our evolved psychologies and cognitive capacities (and by extension the discoveries of science) did not evolve to deliver us truth—only fitness. Even Darwin seems to have been troubled by this, as reflected in a letter to Belfast philosopher William Graham in 1881, not long before Darwin's death: '… with me the horrid doubt always arises whether the convictions of man's mind, which has been developed from the mind of the lower animals, are of any value or at all trustworthy. Would anyone trust in the convictions of a monkey's mind, if there are any convictions in such a mind?' (http://www.darwinproject.ac.uk/entry-13230).

Temporal Discounting and Scope Insensitivity

Evolution has given humans a capacity (unlike any other animal) to anticipate future events that might occur, and to assess, with some degree of accuracy the probability of their occurrence. But because they only 'might occur', evolution has left us generally ill-equipped to respond effectively in advance of these future events. Instead, we have evolved (like many or most other animals) a keen sense of awareness of— or at least readiness needed for effective response to—immediate, clear, near and present problems. This was always of greater importance for promoting gene transmission success. As Penn (2003) put it: 'Squirrels that spend too much of their time preparing and storing food for the upcoming winter, rather than addressing their immediate survival needs (watching out for predators), are less likely to make it to see the winter'. In addition, we also have an evolved motivation for periodic distractions of leisure (Leisure Drive, Chap. 9)—experiences that effectively 'arrest' the mind firmly in the immediate present (because they are pleasurable), thus buffering it from self-impermanence anxiety, but also distracting it from other future disquietudes.

The spatial and temporal scales of the global environmental crisis are generally too large therefore to evoke within us any intrinsic sense of need for urgent response. The ecological crisis is always set for 'tomorrow', as a forecast, and as Burr (2008) put it: '… tomorrow is a slippery idea. It never comes, so it essentially refers to nothing. … As long as the test for an ecological crisis is in the future, we become insured against seeing that we are immersed in it here and now'. There is nothing about having an instinctual motivation to 'save the planet' that would have ever rewarded the reproductive success of our predecessors. Ehrlich and Ornstein (2010) elaborate:

> It seems likely that natural selection long favoured our brain's tendency to habituate—to tune out all but 'big news' recent changes in the world or nearby events. This tendency is still with us. We notice the air conditioner start up, but then its steady hum disappears from consciousness. This process, in which we cancel out ongoing unchanging activity, allows us to focus on sharp and immediate changes, like a sudden appearance of a leopard or a Jaguar sedan driven across our path, but it makes the environmental background seem even more constant than it is.

This points to an unsettling deduction that does not bode well for a biosocial management agenda: effective global response to the converging catastrophes of the twenty-first century may require that they hit us hard, in full force, right on our 'doorsteps', forcing us to literally feel their pain (and hence probably too late to do much of anything about it).[2] This chimes with what Naomi Klein (2007) called the 'shock doctrine': people are predisposed to react only to disasters (and disasters are manufactured as part of a deliberate policy framed by corporations with hidden influence in government). But there is nothing manufactured about the current threats of global disaster for civilization as we know it. The 'shock doctrine' perhaps explains the popularity of 'disaster films', which, according to Dixon (1999), also has another allure that speaks to our intrinsic self-impermanence anxiety: 'People go to disaster movies to prove to themselves that they can go through the worst possible experience but somehow they're immortal'.

---

[2] The famous collapse of civilization on Easter Island in the eighteenth century is a case in point (Diamond 2005).

Aversion for Darwinism

*Man is the result of a purposeless and materialistic process that did not have him in mind.
He was not planned. He is a state of matter, a form of life . . . akin. . . to all of life and indeed
to all that is material.* (Simpson 1949)

Ironically, what is needed most for the very foundation of biosocial management,
may, itself, be the toughest goal to achieve: broader public understanding and acceptance of Darwinism, even to the deepest level (Box 2.2) explored in this book.
Echoing Simpson above, Barash (2000) offers another version of its bleak edict: '...
there is no intrinsic, evolutionary meaning to being alive. We simply are. And so are
our genes. Indeed, we are because of our genes, which are for no other reason than
that their antecedents have avoided being eliminated.' These words—true as they
are—are unlikely to win many converts to Darwinism. From Aarssen (2010):

*The truth of Darwinism itself demands, paradoxically, that the prospect of universal
Darwinism is preposterous—that a future world fully indoctrinated by Darwinism is, in
fact, anti-Darwinian. Such a world would mean that the human species would need to have
evolved to a state where it had no choice but to confront consciousness head-on—with no
credible distractions or delusions available to protect the mind from itself. Foundational
Darwinism, with no patience for delusions, would necessarily applaud this on the principle
of reason, but in so doing undercut its own conclusion that the conscious human mind is
generally designed by Darwinian natural selection to routinely flee from the reasoning of
Darwinism. ... For its history of marginal status (compared with religion at least) as a
cultural worldview, therefore, Darwinism essentially has nothing but itself to blame.
Darwinism is like a snake determined to extend its reach, while feasting on its own tail.*

The Prosocial Imperative

*The imagination, by means of which alone I can anticipate future objects, or be interested
in them, must carry me out of myself into the feelings of others by one and the same process
by which I am thrown forward as it were into my own being, and interested in it.*
(Hazlitt 1805)

Evolution has given humans a lot of things, and one of the good ones is a prosocial nature, defined by an empathic instinct, and dispositions to cooperate effectively within a social group, including with non-kin helping behaviour (Chap. 8).
These features of our human inclinations, more than any other, give reason for optimism that biosocial management can come to our rescue in the twenty-first century—at least in terms of easing the collapse, and preparing the way for a better
model of civilization to unfold in the coming decades. Referring to the human moral
sense, Darwin (1871) wrote: 'As soon as this virtue is honored and practiced by
some few men, it spreads through instruction and example to the young, and eventually becomes incorporated in public opinion'.

Empathy in particular has been the subject of several recent analyses on this
subject (de Wall 2009; Rifkin 2009; Ehrlich and Ornstein 2010; Christakis 2019),
prescribing that hope for a promising future lies in directing biocultural evolution in

ways that will further extend the capacity and impact of our empathic instinct to a global scale. From Ehrlich and Ornstein (2010):

> *Know it or not, all of us are now walking the same tightrope, and we need to keep our balance together. At the circus, our mirror neurons keep us squirming in our chairs, and we are almost overwhelmed with empathy. But now we need to generalize that emotion from circus to globe. We must act as one with those with whom we share the now-swaying civilization tightrope, with millions of fellow citizens of our nations, with seven billion fellow citizens of Earth, and with trillions of Homo sapiens who, if we succeed, will have the opportunity to try to walk the tightrope in the very long-term future. The good news is that we're going with the tide of history. Our human families have been extending and our empathy expanding for thousands of years. For the final step to save civilization, everyone must strive to transform that tide into a tidal wave.*

Importantly, however (as emphasized in Chaps. 2 and 8) empathy and non-kin helping behaviour do not indicate altruism—not in the sense of imposing a genetic fitness cost. And humans tend to be prosocial not just because they discuss it with each other and decide, through social learning, that it would be a nice thing to do. Prosociality is also in our 'selfish' genes—'whispering' to us with 'prepared learning'—inherited from ancestors because their prosocial behaviours contributed in propelling copies of these genes into future generations. And the benevolence in these behaviours was not generally handed out indiscriminately. From Aarssen (2013):

> *It has never been human nature to be generally content with one's state of affairs, because it has always been in the best interests of one's genes to want more for 'Me' and often for 'Us' as well, but while usually being indifferent to—or wanting less for—'Them.' While it seems likely that human nature has indeed been shaped in part by inheritance of an instinctual capacity for empathy toward 'Us'—the members of one's social group—the fitness reward for this may have been normally rooted in tribalism and territorialism. The latter are necessarily accompanied by suspicion or fear of rival groups, representing the evolutionary roots of parochialism, xenophobia, and enthnocentrism.*

There seems to be signs of momentum for identifying social learning strategies that might create a more inclusive 'We' generation to succeed the 'Me' generation (e.g. Greenberg and Weber 2008; Unger 2009). But at the same time, according to other analysts, narcissism has become an epidemic (Twenge and Campbell 2009), and a recent meta-analysis of survey data on the 'Me' generation points to a sharp contemporary decline in empathic concern (Konrath et al. 2001).

We might also wonder whether those who volunteer to help save an over-crowded planet by having only one child or choosing to remain childless might also be those who are most likely to have the genetic bequeathal needed to instil a global-scale target for empathy within future generations. The only thing that empathy can be counted on to save with certainty may be the transmission success of genes that promote empathic behaviour. If this is true, then it may be naïve to count on empathy as a remedy for developing any new model of civilization for our descendants that is based on constrained opportunities for these descendants to—in turn—propel their genes into their descendants (Aarssen 2013). We can only hope that this is not so, but if so, hope that cultural selection will nevertheless advance well enough, perhaps just because humans—through their intrinsic attraction to distracting, 'escape from self' pleasures (Chap. 9)—are likely to be happier when they 'do the right thing'.

Moral Enhancement

But what if—despite instincts for empathy and helpful cooperation with others—
Homo sapiens is just not sufficiently equipped to 'do the right thing', at least not on
a global scale, in response to the current crisis? Garrett Hardin (1968) made a
famous pitch for this warning in his 'Tragedy of the Commons'. Penn (2003)
summarizes:

> *... people are unlikely to conserve common-pool resources when they lack confidence that
> others will show similar restraint. As a resource becomes overexploited, prudent restraint
> only yields opportunity costs, and so users have incentives to get their fair share before it is
> all gone ... Each individual faces a dilemma in which they ask themselves, 'Why should I
> sacrifice and minimize my reproduction and environmental impact if others do not do
> the same?'*

Hardin's (1968) main conclusion was applied especially to overpopulation:
'Ruin is the destination toward which all men rush, each pursuing his own interest
in a society that believes in the freedom on the commons ... No technical solution
can rescue us from the misery of overpopulation. Freedom to breed will bring ruin
to all.'

Moral behaviour, it seems, normally tends to be strategically deployed in various
degrees and at different times depending on the perceived balance of costs and benefits. As Russian novelist and social critic Aleksandr Solzhenitsyn put it (cited in
Wilson 1978):

> *If only it were all so simple! If only there were evil people somewhere insidiously commit-
> ting evil deeds, and it were necessary only to separate them from the rest of us and destroy
> them. But the line dividing good and evil cuts through the heart of every human being. And
> who is willing to destroy a piece of his own heart?*

The strategies that account for the emergence of *Homo sapiens* as the dominant
social animal, as Wilson (2012) put it, '... were written as a complicated mix of
closely calibrated altruism, cooperation, competition, domination, reciprocity,
defection, and deceit'. The cognitive and social skills required to modulate an optimal degree of niceness and helpfulness has served well the best interests of our
ancestors' genes. And having a sharp empathic instinct would have been an important tool in refining and honing those skills.

Accordingly, an important mission for biosocial management will probably
involve implementation of remedies for 'moral enhancement'. Inspiration and education for collective, ethical obligation to the global environment, its biosphere, and
its peoples can come from both religious and secular precepts (Moore and Nelson
2010). Guiding themes are likely to include universal moral codes like the 'Golden
Rule' ('Do unto others as you would have them do unto you'), and a reasonable
balance of widely accepted standards for human rights. This might include, for
example, recognition that individuals have a fundamental right to life, liberty, and
the pursuit of personal happiness—but at the same time, recognition that the best
interests of the individual (e.g. in seeking personal happiness) normally cannot be
regarded as greater than the best interests of the society at large (which will

necessarily include interests for protecting the sustainability of ecosystem services)—including for future generations (more on this below).

Successful moral enhancement will require not just education, but probably also social pressure like public shaming to promote incentives for social/institutional redesign that helps to minimize negative effects of marketing and advertising—e.g. through 'green' imaging, and carbon tax. As Penn (2003) put it:

> *Education provides the information necessary for making individuals aware of their common interests, and it is especially effective for employing shame and other types of social pressure. Because not all people stand to gain equally from conserving natural resources, social pressure and coalitional enforcement may be the only tools that individuals can use to resist manipulation from social dominants. Humans are highly social animals that care about their reputation, and social pressure appears to provide a strong incentive to change behavior. … It is human nature to want more than what is necessary to survive and reproduce—more resources, more social status, more mates—but it is also human nature to want fairness and to shame individuals that behave selfishly!*

The key in this approach probably lies in what we are able to teach youth, e.g. within school curricula. History has shown very often that young children, if carefully guided, can be rather easily 'brain-washed', so to speak, because their cognitive function is plastic—developing in response to social and environmental influence well into the third decade of life (Mitchell 2018). Some of this, at least, could (and should) take advantage of genetic bequeathals like our intrinsic need for self-esteem and a sense of 'belonging', and motivation for 'extension of self' through accomplishment and reputation. Status competition, for example, through 'conspicuous conservation', might be used to promote pro-environmental behaviour: e.g. by 'going green to be seen' (Griskevicius et al. 2010).

Persson and Savulescu (2012) call for moral '*bio*-enhancement'—a more direct approach using pharmaceuticals, supplements, and therapies involving mood-altering 'feel good' hormones like oxytocin and serotonin. Other potential (and more controversial) remedies from biomedicine might include genetic engineering and 'designer babies'. Some researchers today are embracing an even more ambitious goal for 'transhumanism', where—through technologies combining molecular genetics, nanotechnology, robotics, and artificial intelligence in evolving configurations—future humans might theoretically be designed/programmed (using interfacing and replaceable computer chips and circuits) with only desirable traits and behaviours. In its most ambitious form, the 'Singularity' (Kurzweil 2005) promises/predicts a future with universal happiness and immortality. Presumably this would need to include programing that nullifies our anciently intrinsic self-impermanence anxiety, and erases our deeply ingrained motivations for buffering this anxiety through Leisure and Legacy Drives (Chaps. 9 and 10). A recent rumination from E. O. Wilson (2012), however, provides a sombre perspective on these predictions: '… the human condition is an endemic turmoil rooted in the evolution processes that created us. The worst in our nature coexists with the best, and so it will ever be. To scrub it out, if such were possible, would make us less than human'.

The road to transhumanism, if it ever becomes fully realized as an option, may not be a popular route in democratic societies. If transhumanism is viewed as 'less than human', then it may also be regarded as 'other-than-self'. This may appeal to our intrinsic attraction to 'escape-from-self' (Chap. 9), but may be a violation of our intrinsic attraction to 'extension-of-self' (Chap. 10). The current crisis of civilization itself, however, poses a more immediate violation of our intrinsic attraction to self-extension. Choron (1964) described a similar violation imposed by the scare of atomic weapons escalation in the 1960s:

> ... in the face of such a danger, the non-personal ("symbolic") immortalities, which were predicated on the continuation of civilization and of the human race itself, lose their consoling force. The assurances they gave that one would live on in one's progeny or in one's works, in other words that even after death one would not completely 'fall outside' the world, become invalidated when the world itself is bound to disappear.

The Moral Dilemma of the Twenty-First Century

> The Anthropocene raises the vexed question as to whether the seemingly innate process of social development set in train by the evolutionary innovations of fire, language, tool use, and farming engenders an unavoidable clash between any kind of modernity and the ecological integrity of the biosphere. (Zywert and Quilley 2020)

Since the dawn of agriculture, individuals of every large wild species of mammal have left, on average, no more than one descendant—except for one: *Homo sapiens*, together of course with our domesticated livestock and pets. Over the past 12,000 years, biocultural evolution (Box 6.4), shaped by the big four human drives (Box 9.1), has enabled our ancestors to propel copies of their genes into future generations at a pace that has multiplied our numbers by over 1800-fold, and rising still (https://ourworldindata.org/world-population-growth). Now, in the twenty-first century, equipped with keen prosocial and moral inclinations, combined with a perpetual need to manage self-impermanence anxiety—assuaged by Leisure and Legacy Drives—we find ourselves faced with a profound moral dilemma. It obtains from the following premises:

Premise 1: Biocultural evolution spawned the institutions of modern agriculture and medicine/health care, with a shared underlying mission: *a moral obligation to promote and facilitate as many healthy human lives as possible, and lives that can last as long as possible*—by researching and implementing technological innovations and practices designed to treat, prevent, or minimize suffering and limitations from starvation, malnutrition, illness, infection, disease, injury, disability, and infirmity—and by making these technologies available (with influence from economic and political cultures, and of course in some cases, motivations for earning financial profit, status, and prestige) to all of humanity wherever there is need (Box 5.4).

Premise 2: Promoting longer healthier lives reduces mortality rate, and increases birth rate (healthier people can have more babies)—and so *promotes larger population sizes* (Box 12.2, top panel). 'Every increase in food production to feed an increased population is answered by another increase in population' (Burr 2008).

Premise 3: Growing population sizes, and the technologies to support them, *increase the strain on limited planetary resources*—ecosystem services—that are already now being extracted, or destroyed, at rates that vastly exceed their available supplies and renewal rates.

Premise 4: Destruction of ecosystem services *harms the opportunity for future generations (the descendants of all alive today) to enjoy long, healthy lives*.

Premise 5: It is *wrong to compromise the health and longevity of individuals living in future generations*.

Conclusion: *We have a moral obligation to avert that harm*.

Therein presents our dilemma: how do we reconcile these two conflicting moral obligations? For as long as we have been able to do so—through innovations of technology and supporting industries in agriculture and medicine/health care—we have assumed a global-scale moral obligation to prolong human life, by providing people in need with adequate food and enabling them to live healthier into older age. But in doing this now—with growing knowledge and acceptance (e.g. Lawton 2019a) of the perils of over-population and its associated converging environmental catastrophes of the twenty-first century (Chap. 1)—we violate our moral obligation not to imperil these opportunities for future generations (which include our own children and other living kin).

Note that the moral dilemma here is rooted in an evolved psychology informing a sense of obligation to the welfare of others, but one that instinctively wanes with decreasing genetic relatedness (Chap. 6)—and, as discussed in Chap. 2, one's genetic relatedness to individual descendants necessarily falls off sharply with each successive generation. Accordingly, we are much more intrinsically inclined to ensure that ourselves and our living descendants (as opposed to future descendants) have abundant access to products and services from agriculture and medicine/health care. At the same time, however, as Krznaric (2020) put it, 'Future generations deserve good ancestors … we can rest assured that there is one thing that our descendants will want to inherit from us: a living world in which they can survive and thrive'. Species that destroy the ecosystems that their descendants will need in order to leave descendants go extinct—unless they evolve in time to become less destructive, or to need different ecosystems (perhaps as a new species). An apparent alternative option under consideration—and anticipated by some wealthy *Homo sapiens*—is to abandon our beleaguered planet and purchase access to celestial ecosystems, together with a seat on a rocket to the moon or Mars (Mehta 2017; Zorthian 2017; Rushkoff 2018).

Our biocultural evolution has largely disconnected humanity from its moral obligation to the future on planet earth. We seem utterly blind to the harmful, long-term consequences of modern agriculture, where research agendas are relentlessly committed to finding new technologies and practices for feeding more and more people. These include, for example, adjusting crop varieties and increasing cropping efficiency to maximize yields (e.g. http://www.fao.org/wsfs/forum2050/wsfs-forum/en/, http://sri.ciifad.cornell.edu/), including through closing 'yield gaps' on underperforming lands (Foley et al. 2011), improving water management and increasing nitrogen and phosphorus use efficiency (https://eatforum.org/eat-lancet-commission/), and 'precision agriculture' (https://croplife.ca/field-notes-precision-agriculture-canada/); developing higher-yielding and more pest resistant crops, e.g. involving 'speed breeding' (Watson et al. 2018), CRISPR gene editing (Le Page 2018a), and engineered crops with a 'turbocharged' version of photosynthesis (Fedunik 2020); boosting animal nutrient use efficiency (Gu et al. 2019); and developing high-rise (vertical) farming (Vaughan 2019), and 'cellular agriculture'—the production of animal-sourced foods from cell culture (https://new-harvest.org/).

Similarly, the research agendas of modern medicine and health care are relentlessly committed to finding new technologies, products, and practices that enable more and more people to live longer, healthy lives (with greater life expectancy). These include new diagnostics, cures, pharmaceuticals, and other treatments for disease (National Institutes of Health 1997), including CRISPR gene editing (Le Page 2018b; Ledford 2020); placing patients in suspended animation (emergency preservation and resuscitation (EPR)) making it possible to fix traumatic injuries that would otherwise cause death (Thomson 2019); new technologies, remedies, and interventions (e.g. drugs, implants, cloning, tissue regeneration, stem cell technology), for slowing the pace of aging (and maintaining health during it) (e.g. Hamzelou 2015; Lawton 2019b, c, https://www.theage.com.au/interactive/2015/never-say-die/), including with integration of biomarkers and artificial intelligence (https://deeplongevity.com/); and (more controversially) using cryogenics—preserving and deep-freezing an ailing body with the intention of having it revived when future technology has provided a cure (Buder 2019)—and (even more controversially) future technology for creating a digital copy of one's mind in a computer ('mind-uploading') (Laakasuo et al. 2018), and presumably incorporating that eventually into a robotic/cyborg body, all (theoretically) repairable indefinitely with hardware replacement parts and programming upgrades (Veldhuis and Thomas 2019).

Our contemporary 'March of Progress'—with its aims to feed the world, end disease, and extend healthy human life spans—is an injustice to the future of humankind. If not aggressively combined with birth limitation on a global scale (Hardin 1968; Greguš and Guillebaud 2020)—plus (much more difficult) placing some limits on the reduction in death rate—it will promote escalation of unspeakable misery for our descendants on a more crowded, and more impoverished planet. Again, as Ehrlich (1968) put it, 'Whatever your cause, it will be a lost cause without population control'. And to this I would add: It would also be a lost cause without Biosocial Management proposed in the present chapter—where 'becoming the solution' prioritizes not just identifying the corrections needed to ensure that we can have healthy descendants living on a healthy planet. Success for the latter will also require a new cultural revolution based a deeper and more broadly public understanding of the evolutionary roots of how and why we arrived at this predicament—as products of our own evolved psychology, involving essential needs shaped by the Big Four Human Drives (Box 9.1). This will require a collective empathy for these needs, and prescriptions for how they can be met effectively in ways that equip us to respond appropriately to our contemporary moral obligations, and at the same time, promote and successfully deliver the restored and sustainable ecosystem services that future generations are entitled to.

In the next (final) chapter, we will examine our greatest challenge in confronting what we are, and how we must urgently address it through deployment of Biosocial Management in launching this new cultural revolution.

In modern health systems … the rights of individuals overshadow other facets of population or planetary health. This was a progressive and functional approach in the Holocene, but it is conceivable that other units of analysis (populations of humans, microbes, animals, plants, and/or their interactions, for instance) could rise to prominence within health systems developing in a very different kind of modernity. … imagine, for instance, how individual rights could be affected if the health of ecosystems was valued above the health of individual humans. (Zywert and Quilley 2020)

References

Aarssen LW (2005) Why is fertility lower in wealthier countries? the role of relaxed fertility-selection. Popul Dev Rev 31:113–126

Aarssen LW (2010) Darwinism and meaning. Biol Theory 5:296–311

Aarssen LW (2013) Will empathy save us? Biol Theory 7:211–217

Alexander RD (1971) The search for an evolutionary philosophy of man. Proc Roy Soc Victoria 84:99–120

Australian Broadcasting Corporation, Computer predicts the end of civilization (1973) Aired 9 Nov 1973. https://aeon.co/videos/civilisation-peaked-in-1940-and-will-collapse-by-2040-the-data-based-predictions-of-1973

Barash DP (2000) Evolutionary existentialism, sociobiology, and the meaning of life. Bioscience 50:1012–1017

Barash DP (2008) Natural selections: selfish altruists, honest liars, and other realities of evolution. Bellevue Literary Press, New York

Barnosky AD, Haely EA, Bascompte J et al (2012) Approaching a state shift in earth's biosphere. Nature 486:52–58

Baumeister RF (2010) Is there anything good about men. Oxford University Press, New York

Bloom DE, Canning D, Lubet A (2015) Global population aging: facts, challenges, solutions & perspectives. Daedalus 144:81–92

Bolund E, Hayward A, Pettay JE, Lummaa V (2015) Effects of the demographic transition on the genetic variances and covariances of human life-history traits. Evolution 69:747–755

Bradshaw CJA, Brook BW (2014) Human population reduction is not a quick fix for environmental problems. Proc Natl Acad Sci 111:16610–16615

Braumoeller B (2013) Is war disappearing? Annual Meeting of the American Political Science Association, Chicago. Available at SSRN: http://ssrn.com/abstract=2317269

Bribiescas RG (2011) An evolutionary and life history perspective on the role of males on human futures. Futures 32:729–739

Bribiescas RG (2021) Evolutionary and life history insights into masculinity and warfare: opportunities and limitations. Curr Anthropol 62(S23):S38–S53

Brickman P, Campbell D (1971) Hedonic relativism and planning the good society. In: Apley MH (ed) Adaptation level theory: a symposium. Academic Press, New York, pp 287–302

Brito L, Stafford-Smith M (2012) State of the planet declaration. Planet under pressure: New knowledge towards solutions conference, London, 26–29 Mar 2012. Available at: http://www.

igbp.net/download/18.6b007aff13cb59eff6411bbc/1376383161076/SotP_declaration-A5-for_web.pdf

Buder E (2019) Die. Freeze body. Store Revive. The Atlantic, 20 June 2019. https://www.theatlantic.com/video/index/591979/cryonics/

Burr C (2008) Culture quake: your children's real future. Trafford Publishing, Bloomington

Butler T, Kanyoro M, Ryerson WN, Crist E (2015) Overdevelopment, overpopulation, overshoot. Goff Books, New York

Cafaro P, Crist E (2012) Life on the brink: environmentalists confront overpopulation. University of Georgia Press, Athens

Carroll L (1872) Through the looking glass. Macmillan, London

Choron J (1964) Modern man and mortality. The Macmillan Company, New York

Christakis NA (2019) Blueprint: the evolutionary origins of a good society. Little, Brown Spark, New York

Cochran G, Harpending H (2009) The 10,000 year explosion: how civilization accelerated human evolution. Basic Books, New York

da Silva KB (2012) Evolution-centered teaching of biology. Annu Rev Genomics Hum Genet 13:363–380

Darwin CR (1871) The descent of man, and selection in relation to sex. John Murray, London

Darwin CG (1953) The next million years. Doubleday, New York

De Duve C (2010) Genetics of original sin: the impact of natural selection on the future of humanity. Yale University Press, New Haven

De Wall F (2009) Age of empathy: lessons for a kinder society. Harmony Books, New York

Diamond J (1992) The third chimpanzee: the evolution and future of the human animal. Harper, New York

Diamond J (2005) Collapse: how societies choose to fail or succeed. Penguin Group, New York

Diaz S, Settele J, Brondizio ES et al (2019) Pervasive human-driven decline of life on earth points to the need for transformative change. Science 366:eaax3100. https://science.sciencemag.org/content/366/6471/eaax3100/tab-pdf

Dilworth C (2010) Too smart for our own good: the ecological predicament of human kind. Cambridge University Press, Cambridge

Dixon WW (1999) Disaster and memory: celebrity culture and the crisis of hollywood cinema. Columbia University Press, New York

Dobzhansky T (1962) Mankind evolving: the evolution of the human species. Yale University Press, New Haven

Dobzhansky T (1973) Ethics and values in biological and cultural evolution. Zygon 8:261–281

Dobzhansky T, Allen G (1956) Does natural selection continue to operate in modern man? Am Anthropol 58:592–604

Dyer G (2008) Climate wars. Random House Canada, Toronto

Easterlin RA, McVey LA, Switek M (2010) The happiness-income paradox revisited. Proc Natl Acad Sci 107:22463–22468

Ehrlich P (1968) The population bomb. Ballantine Books, New York

Ehrlich P, Ehrlich A (2008) The problem is simple: too many people, too much stuff. Yale Enviro 360, 4 Aug 2008. Yale School of the Environment. https://e360.yale.edu/features/too_many_people_too_much_consumption

Ehrlich PR, Ehrlich AH (2013) Can a collapse of global civilization be avoided? Proc R Soc B 280:20122845

Ehrlich PR, Ornstein RE (2010) Humanity on a tightrope: thoughts on empathy, family, and big changes for a viable future. Rowman and Littlefield, Plymouth

Eiseley L (1957) The immense journey. Random House, New York

Engelman R (2010) More: population, nature, and what women want. Island Press, Washington DC

Fedunik L (2020) The secret power of the tequila plant that could help feed the world. New Sci, 15 July 2020. https://www.newscientist.com/article/mg24732911-400-the-secret-power-of-the-tequila-plant-that-could-help-feed-the-world/

Foley JA et al (2011) Solutions for a cultivated planet. Nature 478:337–342

Gluckman P, Hanson M (2019) Ingenious: the unintended consequences of human innovation. Harvard University Press, Cambridge

Greenberg EH, Weber K (2008) Generation we: how millennial youth are taking over America and changing our world forever. Pachatusan, Emeryville

Greguš G, Guillebaud J (2020) Doctors and overpopulation 48 years later: a second notice. Eur J Contracept Reprod Health Care 25:409–416

Griskevicius V, Van den Bergh B, Tybur JM (2010) Going green to be seen; status, reputation, and conspicuous conservation. J Pers Soc Psychol 98:392–404

Gu B, Zhang X, Bai X, Fu B, Chen D (2019) Four steps to food security for swelling cities. Nature 566:31–33

Hamzelou J (2015) Secret to old-age health could lie in purging worn-out cells. New Sci, 16 Sep 2015. https://www.newscientist.com/article/mg22730390-500-secret-to-old-age-health-could-lie-in-purging-worn-out-cells/

Hardin G (1968) The tragedy of the commons. Science 162:1243–1248

Hardin G (1999) The Ostrich factor: our population Myopia. Oxford University Press, New York

Harper JL (1977) Population Biology of plants. Academic Press, London

Harrison M, Wolf N (2012) The frequency of wars. Econ Hist Rev 65:1055–1076

Haselton MG, Nettle D, Murray DR (2016) The evolution of cognitive bias. In: Buss D (ed) Handbook of evolutionary psychology, 2nd edn. John Wiley & Sons, New York

Hayden EC (2015) Ageing pushed as treatable condition. Nature 522:265–266

Hazlitt W (1805) An essay on the principles of human action. J Johnson, St. Paul's Churchyard, London. https://archive.org/details/anessayonprinci00hazlgoog

Heinberg R (2011) The end of growth: adapting to our new economic reality. Gabriola Publishers, BC Canada

Homer-Dixon T (2007) The upside of down: catastrophe, creativity and the renewal of civilization. Vintage Canada, Toronto

Idiocracy (2006). https://en.wikipedia.org/wiki/Idiocracy

Joyce K (2010) Quiverfull: inside the Christian patriarchy movement. Beacon Press, Boston

Kahan DM, Jenkins-Smith H, Braman D (2010) Cultural cognition of scientific concensus. J Risk Res 14:147–174

Kanazawa S (2014) Intelligence and childlessness. Soc Sci Res 48:157–170

Khazan O (2019) I was never taught where humans came from: many American students, myself included, never learn the human part of evolution. The Atlantic, 19 Sep 2019. https://www.theatlantic.com/education/archive/2019/09/schools-still-dont-teach-evolution/598312/

Kingsolver B (1995) High tide in Tucson. Harper Collins, New York

Klein N (2007) The shock doctrine: the rise of disaster capitalism. Henry Holt & Co., New York

Konrath SH, O'Brien EH, Hsing C (2001) Changes in dispositional empathy in American college students over time: a meta-analysis. Personal Soc Psychol Rev 15:180–198

Krznaric R (2020) Future generations deserve good ancestors. Will you be one? Psyche, 21 July 2020. https://psyche.co/ideas/future-generations-deserve-good-ancestors-will-you-be-one

Kurzweil R (2005) The singularity is near: when humans transcend biology. Penguin, New York

Laakasuo M, Drosinou M, Koverola M et al (2018) What makes people approve or condemn mind upload technology? Untangling the effects of sexual disgust, purity and science fiction familiarity. Palgrave Communicat 4:84. https://doi.org/10.1057/s41599-018-0124-6

Lawton G (2019a) We need to talk about how population growth is harming the planet. New Sci, 22 May 2019. https://www.newscientist.com/article/mg24232310-100-we-need-to-talk-about-how-population-growth-is-harming-the-planet/

Lawton G (2019b) Anti-ageing drugs are coming that could keep you healthier for longer. New Sci, 24 Apr 2019. https://www.newscientist.com/article/mg24232270-100-anti-ageing-drugs-are-coming-that-could-keep-you-healthier-for-longer/

Lawton G (2019c) Can a supplement slow the natural processes of ageing? New Sci, 30 July 2019. https://www.newscientist.com/article/2211927-exclusive-can-a-supplement-slow-the-natural-processes-of-ageing/

Laybourn-Langton L, Rankin L, Baxter D (2019) This is a crisis: facing up to the age of environmental breakdown. Institute for Public Policy Research, London. https://www.ippr.org/research/publications/age-of-environmental-breakdown

Le Bon G ([1896] 2001) The crowd: a study of the popular mind. Batoche Press, Kitchener, Canada. http://socserv.mcmaster.ca/econ/ugcm/3ll3/lebon/Crowds.pdf

Le Page M (2018a) The second great battle for the future of our food is underway. New Scientist, 4 July 2018. https://www.newscientist.com/article/mg23931852-500-the-second-great-battle-for-the-future-of-our-food-is-underway/

Le Page M (2018b) DNA editing before birth could one day massively expand lifespans. New Sci, 30 Aug 2018. https://www.newscientist.com/article/2178313-dna-editing-before-birth-could-one-day-massively-expand-lifespans/

Ledford H (2020) Quest to use CRISPR against disease gains ground. Nature 577:156

Lidicker WZ (2020) A Scientist's warning to humanity on human population growth. Global Ecol Conservat 24:e01232

Malthus TR (1798) An essay on the principle of population as it affects the future improvement of society with remarks on the speculations of Mr. In: Godwin M (ed) Condorcet and other writers, 1st edn. Johnson in St Paul's Church-yard, London

Martin K, Sauerborn J (2013) Agroecology. Springer, Dordrecht

McKenna M (2013) Antibiotic resistance: The last resort. Nature 499:394–396

Mehta J (2017) How to colonize the moon? Medium, 28 Nov 2017. https://medium.com/teamindus/how-to-colonize-the-moon-6723a5c388de

Meyer-Abich KM (1997) Humans in nature: toward a physiocentric philosophy. In: Ausubel JH, Langford HD (eds) Technological trajectories and the human environment. National Academy Press, Washington, DC, pp 168–184

Miller G (2007) Runaway consumerism explains the fermi paradox. In: Brockman J (ed) What is your dangerous idea? Harper Perennial, New York, pp 240–243

Mitchell KJ (2018) Innate: how the wiring of our brains shapes who we are. Princeton University Press, Princeton

Mitzberg H (2014) Rebalancing society: radical renewal beyond left, right, and center. Berrett-Koehler Publishers, Oakland

Moore KD, Nelson MP (2010) Moral ground: ethical action for a planet in peril. Trinity University Press, San Antonio

Morello L (2013) Warming climate drives human conflict. Nature. http://www.nature.com/news/warming-climate-drives-human-conflict-1.13464

National Institutes of Health (1997) Setting Research Priorities at the National, Institutes of Health. National Institutes of Health, Bethesda, MD

Nature (2013) Plant biotechnology: Tarnished promise. Nature 497:21. (2013). https://doi.org/10.1038/497021a

Nekola JC, Allen CD, Brown JH et al (2013) The Malthusian–Darwinian dynamic and the trajectory of civilization. Trends Ecol Evol 28:127–130

Nettle D (2005) Happiness: The Science Behind Your Smile. Oxford University Press, Oxford

New Scientist, Special report: how our economy is killing the earth. New Sci (2678), 16 Oct 2008. http://www.newscientist.com/article/mg20026786.000-special-report-how-our-economy-is-killing-the-earth.html?full=true

Nuwer R (2012) Are humans getting intellectually and emotionally stupid? Smithsonian Magazine, 14 Nov 2012. https://www.smithsonianmag.com/smart-news/are-humans-getting-intellectually-and-emotionally-stupid-121924658/?no-ist

Penn DJ (2003) The evolutionary roots of our environmental problems: toward a Darwinian ecology. Q Rev Biol 78:275–301

Persson I, Savulescu J (2012) Unfit for the future: the need for moral enhancement. Oxford University Press, New York

Pimentel D, Whitecraft M, Scott ZR et al (2010) Will limited land, water, and energy control human population numbers in the future? Hum Ecol 38:599–611

Rifkin J (2009) The empathic civilization: the race to global consciousness in a world in crisis. Polity Press, Cambridge

Ripple WJ, Wolf C, Newsome TM, Barnard P, Moomaw WR (2020) World scientists' warning of a climate emergency. Bioscience 70:8–12

Roberts SC (ed) (2012) Applied evolutionary psychology. Oxford University Press, New York

Rushkoff D (2018) Survival of the richest: The wealthy are plotting to leave us behind. Medium, 5 July 2018. https://onezero.medium.com/survival-of-the-richest-9ef6cddd0cc1

Saad G (2007) The evolutionary bases of consumption. Lawrence Erlbaum, Mahwah, NJ

Saad G (ed) (2011) Evolutionary psychology in the business sciences. Springer, Heidelberg

Schor JB (2010) Plenitude: the new economics of true wealth. Penguin Books, New York. (See also: https://newdream.org/videos/plenitude)

Simpson GG (1949) The meaning of evolution. Yale University Press, New Haven

Smith DL (2007) The most dangerous animal: human nature and the origins of war. St. Martin's Press, New York

Spellberg B (2009) Rising plague: the global threat from deadly bacteria and our dwindling arsenal to fight them. Prometheus Books, New York

Speth JG (2008) The bridge at the edge of the world: capitalism, the environment, and crossing from crisis to sustainability. Yale University Press, New Haven

Stamos DN (2008) Evolution and the big questions. Blackwell, Malden MA

Stearns SC, Byars SG, Govindaraju DR, Ewbank D (2010) Measuring selection in contemporary human populations. Nat Rev Genet 11:611–622

Szent-Györgi A (1970) The Crazy Ape. Philosophical Library, New York

The economist, editorial: vertical farming: does it really stack up? (2010). http://www.economist.com/node/17647627

Thomson H (2019) Humans placed in suspended animation for the first time. New Sci, 20 Nov 2019. https://www.newscientist.com/article/2224004-exclusive-humans-placed-in-suspended-animation-for-the-first-time/

Tollefson J (2018) IPCC says limiting global warming to 1.5 °C will require drastic action: Humanity has a limited window in which it can hope to avoid the worst effects of climate change, according to climate report. Nature 562:172–173

Tollefson J, Gilbert N (2012) Rio report card. Nature 486:20–23

Twenge JM, Campbell WK (2009) The narcissism epidemic: living in the age of entitlement. Free Press, New York

Unger M (2009) We generation: raising socially responsible kids. McClelland and Steward, Toronto

United Nations (2019) Department of Economic and Social Affairs, Population Division. World Population Prospects, 2019. https://population.un.org/wpp/Publications/Files/WPP2019_Highlights.pdf

United Nations Environment Programme (2020) Emissions Gap Report 2020, Nairobi. https://www.unep.org/emissions-gap-report-2020

Updike J (2000) The tried and the treowe. In: Due considerations: essays and criticism (2007). Randon House, New York

Vaughan A (2019) High-tech vertical farming is on the rise – but is it any greener? New Sci, 14 June 2019. https://www.newscientist.com/article/2206633-high-tech-vertical-farming-is-on-the-rise-but-is-it-any-greener/

Veldhuis D, Thomas MG (2019) Your body as part machine. Sapiens, 15 Nov 2019. https://www.sapiens.org/column/machinations/cyborg-future/

Vollset SE, Goren E, Yuan C-W et al (2020) Fertility, mortality, migration, and population scenarios for 195 countries and territories from 2017 to 2100: a forecasting analysis for the global burden of disease study. Lancet 396:1285–1306

Wallace-Wells D (2019) The uninhabitable earth: life after warming. Tim Duggan Books, New York

Washington H (2020) Why do society and academia ignore the 'scientists warning to humanity' on population? J Futures Studies 25:93–106

Watson et al (2018) Speed breeding is a powerful tool to accelerate crop research and breeding. Nat Plants 4:23–29

Wilber N, Withers M (2015) Teaching practices and views of evolution instructors at post-secondary institutions. Evolut Edu Outreach 8:12. http://www.evolution-outreach.com/content/pdf/s12052-015-0038-3.pdf

Wiles JR (2006) Evolution in school's: where's Canada? Canadian Education Association. https://www.edcan.ca/articles/evolution-in-schools-wheres-canada/

Williams T (1955) Cat on a hot tin roof. Penguin Putnam, New York

Wilson EO (1978) On human nature. Harvard University Press, Cambridge

Wilson EO (2012) The social conquest of earth. Liveright Publishing Corporation, New York

World Health Organization, Antimicrobial Resistance: Global Report on Surveillance (2014). http://apps.who.int/iris/bitstream/10665/112642/1/9789241564748_eng.pdf

Wrangham R, Peterson D (1996) Demonic males: apes and the origin of human violence. Houghton Mifflin, New York

Wright R (2019) Can we still dodge the progress trap? The Tyee, 20 Sep 2019. https://thetyee.ca/Analysis/2019/09/20/Ronald-Wright-Can-We-Dodge-Progress-Trap/

Zorthian J (2017) Stephen hawking says humans have 100 years to move to another planet. Time, 4 May 2017. https://time.com/4767595/stephen-hawking-100-years-new-planet/

Zywert K, Quilley S (2020) Health in the anthropocene: living well on a finite planet. University of Toronto Press, Toronto

Chapter 13
Troubled Minds on Runaway Selection

Dream as if you'll live forever. Live as if you'll die today. — James Dean

Irish elk — by Doflein, F. & Hesse, R. (1910) Tierbau und tierleben in ihrem zusammenhang betrachtet/Wikimedia Commons/Public Domain

Evolutionists have used the concept of 'runaway selection' to interpret a number of biological traits considered to have possible connections in driving certain

Some parts of this chapter are reproduced (with permission) from: Aarssen L. (2018) Meet *Homo absurdus*—the only creature that refuses to be what it is. Ideas in Ecology and Evolution. 11: 90–95.

185

species to extinction. One of the most familiar of these speculations involves the Irish elk—a giant deer species, famous for its massive antlers, that went extinct about 8000 years ago (Moen et al. 1999). Antlers are produced in many mammals and are found mostly in males, where they are used as weapons for fighting (or intimidation) in competing for access to females and control of harems. Larger antler size advertises a formidable potential adversary to rival males, but it might also (as with the famous peacock's tail) be associated with an evolved preference in female mate choice (i.e. sexual selection) if antler size is heritable and correlated with male quality. A larger antler is more costly, and so a male that can successfully support and display one may not only deter less endowed males from initiating a challenge; he is also likely to have exceptional health and superior survival prospects (regardless of any advantage in doing battle with other males). According to a popular hypothesis therefore, Irish elk females who produced the highest quality offspring (and who therefore left the most descendants) were generally those with heritable attraction to (and who hence mated with) males displaying these larger ornaments (without knowing—or needing to know—why larger ones were more attractive).

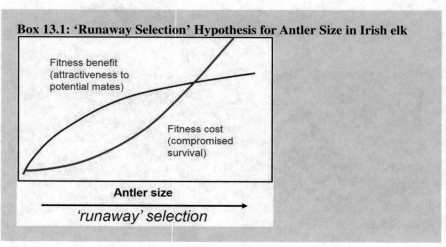

Box 13.1: 'Runaway Selection' Hypothesis for Antler Size in Irish elk

Fitness benefit (attractiveness to potential mates)

Fitness cost (compromised survival)

Antler size

'runaway' selection

Under an assumption of 'runaway selection', therefore, bigger is always better for attracting females, but the evolution of ever larger antlers in subsequent generations starts to take a greater toll on survival success—they just get too big!—thus imposing a higher fitness cost. This eventually increases at a faster rate (per unit increase in antler size) than the rise in attractiveness to potential mates (Box 13.1). If the fitness cost of larger antler size then 'overshoots' its fitness benefit, then larger antlers no longer truthfully signal superior survival success (or superior competitive ability against other males). Extinction risk thus rises sharply—contributing possibly to the demise of the Irish elk.

A similar kind of 'runaway selection', I suggest, may also be described in terms of the effects of self-impermanence anxiety on our evolved psychology. Our lives resemble the lives of chimpanzees more than any other animal. They have a rudimental theory of mind and capacity for culture through social learning. But as Sterelny (2012) put it, chimps '...live in a world as they find it', while humans '...live in a world as they make it'. And, as examined in this book, we have made it mostly a world of accumulating delusions for chasing legacy, and a world of accumulating distractions for chasing leisure in a global culture of addiction to endless (runaway) economic growth. And from this we have made a world of environmental catastrophes that have gone mostly unnoticed by the general public (Chap. 1). The most intelligent animal that has ever lived has an evolved psychology that makes it largely indifferent to the self-destructive consequences of its impact in trashing the planet, demolishing ecosystem services, and annihilating other species and their habitats at a rate 1000 times faster than the natural (historical) rate of extinction.

Legacy Drive and Leisure Drive are both represented in the well-known adage: 'Dream as if you'll live forever; live as if you'll die today'. These self-impermanence anxiety buffers have enabled human minds to function mostly (or at least regularly) unhindered by troubling thoughts of possibly losing one's 'mental life'—and hence minds thinking only infrequently about death (loss of 'material life'). By helping to mitigate the 'curse of self-awareness', selection for strong Legacy and Leisure Drives thus rewarded ancestral gene transmission success, and the descendants became *Homo absurdus*—spending most of our lives trying to convince ourselves that our existence is not absurd—routinely with success, by deploying various 'extension of self' delusions (Chap. 10) and 'escape from self' distractions (Chaps. 9 and 11).

Our ancestors, however, were never really perfectly or permanently convinced, and neither are we; self-impermanence anxiety has always lurked stubbornly beneath the surface (Aarssen 2019). Human motivations did not evolve to deliver us untroubled minds—only fitness. Importantly, *it was the striving for an untroubled mind that delivered fitness*, and this has generated ideal conditions for a kind of escalating 'runaway selection' for minds that become more and more acutely afflicted with self-impermanence anxiety, and hence selection for its mitigation through ever stronger Legacy and Leisure Drives—thus shaping the biocultural evolution that has fueled the frenzied and relentless 'march of progress' that we call civilization (Chap. 5). In other words, a perpetual undercurrent of self-impermanence anxiety has always, more often than not, been favoured by natural selection. But a tipping point is now approaching.

Box 13.2: 'Runaway Selection' Hypothesis for Legacy and Leisure Drives

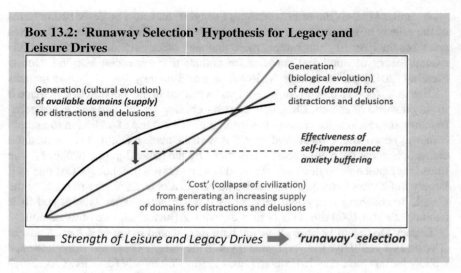

Generation (cultural evolution) of **available domains (supply)** for distractions and delusions

Generation (biological evolution) of **need (demand)** for distractions and delusions

Effectiveness of self-impermanence anxiety buffering

'Cost' (collapse of civilization) from generating an increasing supply of domains for distractions and delusions

━━ *Strength of Leisure and Legacy Drives* ➡ *'runaway' selection*

Runaway selection for Legacy and Leisure Drives has generated two dire conse-quences for humanity in the twenty-first century: (i) a civilization now on the verge of collapse (Chap. 1) and (ii) the ramped-up demands of these drives (resulting from escalating natural selection) are starting to exceed the supply rate of available domains (generated by cultural selection) for satisfying them (Box 13.2). In other words, the effective 'shelf lives' for newer and more convincing cultural delusions (for extension of self) and distractions (for escape from self) become shorter and shorter, while Legacy and Leisure Drives become stronger and stronger. As these intensities accelerate, civilization moves faster and faster to collapse, while we work harder and harder in chasing legacy through accomplishment (Thompson 2019; Wolfe 2019; Brooks 2020; McCallum 2021), with 'all work and no play'—often neglecting leisure (Mustafa 2020)—and while the popularity of legacy through reli-gion wanes (Pew Research Center 2012, Allen 2019). Consequently, incidences of mental health disorders—anxiety, depression, insanity, and suicide—escalate (Curtin et al. 2016; Nock 2016; Horowitz 2018; Whyte 2019) as it becomes harder and harder for cultural products and institutions to provide adequate buffering to keep up with and meet an ever-increasing demand for delusions and distractions, including distractions to buffer the mounting 'eco-anxiety' from living in a collaps-ing civilization (Nugent 2019). Worldwide, close to 800,000 people die by suicide every year—more deaths than are caused by war and homicide combined—and for each death, there are more than 20 suicide attempts (United Nations 2019).

The human condition, as neuroscience has shown us, is essentially hardwired to be an 'insatiable wanting machine' (Fleming 2015). And the biocultural evolution that made us into this has now trapped our civilization on an over-crowded, angst-ridden 'runaway train' (Box 5.4). As Alexander (2021) put it, 'A society might go insane without being aware of its own degeneration. … If profits and economic growth are the benchmarks of success in a society, it simply may not be profitable to expose a society as insane, and even members of an insane society may sooner choose wilful blindness than look too deeply into the subconscious of their own culture'. It's an intriguing irony to consider the possibility/probability that our predecessors who left the most descendants were at least mildly insane.

Striving for Calm

> All of humanity's problems stem from man's inability to sit quietly in a room alone. —
> Blaise Pascal (1670), Pensées

Because humans are a product of the purposeless action of natural selection, there is no grand purpose of human existence; it has no more 'meaning' in the cosmos than does the existence of flies. There is no equanimity to be found in this. It rattles us because *Homo sapiens* evolved a mind for which the meaning of one's life is the most urgent of all questions—a mind that spends its whole life trying to convince itself that its existence is not absurd. Legacy drive and Leisure drive can only be tamed, not dismissed. From Albert Camus (1942):

> I see many people die because they judge that life is not worth living. I see others paradoxically getting killed for the ideas or illusions that give them a reason for living (what is called a reason for living is also an excellent reason for dying). I therefore conclude that the meaning of life is the most urgent of questions.

Similar conclusions have been reached by many other great thinkers from history. Carl Jung (1939, *Wirklichkeit der Seele* — quoted from Choron 1964) wrote that when:

> ... an old man experiences a secret shudder, nay even death fear at the thought that his reasonable life expectancy now amounts to only so and so many years, one is painfully reminded of certain emotions in one's own breast: one looks the other way and tries to change the subject ..." But "When one is alone and it is night and so dark and quiet that one does not see or hear anything but the thoughts that add and subtract the years of one's life, and the long sequence of those unpleasant facts, which prove cruelly how far the hand of the clock has advanced, and the slow and uncheckable approach of that dark wall, which threatens to swallow up irretrievably all I love, desire, possess, hope, and strive for, then all the wise dicta go into hiding and fear descends upon the sleepless like a choking blanket.

And from Bertrand Russell (1957):

> That man is the product of causes which had no prevision of the end they were achieving; that his origin, his growth, his hopes and fears, his loves and his beliefs, are but the outcome of accidental collocations of atoms; that no fire, no heroism, no intensity of thought and feeling, can preserve an individual life beyond the grave; that all the labor of the ages, all the devotion, all the inspiration, all the noonday brightness of human genius is destined to extinction in the vast death of the solar system, and that the whole temple of Man's achievement must inevitably be buried beneath the debris of a universe in ruins — all these things, if not quite beyond dispute, are yet so nearly certain, that no philosophy which rejects them can hope to stand.

The dilemma for humanity today is that it now faces another question that, one would think, should be at least as urgent as the meaning of life: *How can we and our children and our grandchildren survive the converging catastrophes of the twenty-first century?* And yet, the general public seems largely unaware of it, or apathetic (or insane). The dilemma is especially precarious because a root cause of these converging catastrophes—a culture of addiction to consumerism and endless economic and population growth, supported by rapaciously escalating rates of resource extraction and energy production—is the very same culture that evolved to support the deeply ingrained motivational domains of 'working hard' (legacy) and 'playing hard' (leisure) that humans have deployed to convince themselves that their existence is not absurd.

In the way forward, therefore, it will not be enough to focus only on the deployment of available remedies that we already know would address the now urgent question of surviving collapse. This would run the risk of compromising our opportunities to effectively deploy our intrinsic drives for Legacy and Leisure. And we cannot let it happen; people will not be inclined to respond effectively to the crisis of a collapsing civilization unless they believe that their lives can nevertheless have meaning—even if the basis of meaning is in delusion (Chap. 10) and/or distraction (Chap. 9). Legacy and Leisure Drives will need careful management; we ignore them at our peril. The 'work hard–play hard' manifesto inspires today (Aarssen and Crimi 2016) no less than it did two centuries ago:

> Whatever is done, it should be habitually done with earnestness; in every pursuit, exertion should be employed; work hard and play hard; always recollecting that quiescence, the stillness of inactivity is destructive to the mental welfare, and approaches very nearly to the winter of the faculties, the torpor of an hibernating animal, the unprotected state of sleep, or the complete cessation of life (Newnham 1827).

Returning to the interpretation from Albert Camus (1942), 'The meaning of life is the most urgent of questions …'—but the generic answer is simple: whatever keeps one from ending it.

Meaning Without Riding a Runaway Train

> If we merely wish to continue on the scene to indulge our headstrong humours and tormenting passions, we had better begone at once: and if we only cherish a fondness for existence according to the good we derive from it, the pang we feel at parting with it will not be very severe! — William Hazlitt (1822).

The worrying, multi-tasking, auto-pilot minds of our distant ancestors equipped them well for finding, monitoring, and managing the many variable and changing environmental and social resources needed in supporting their primitive drives for Survival and Reproduction. These minds became our minds, but with a regime of 'mind chatter' less troubled by success with survival and reproduction, and obsessed instead mostly with establishing and preserving 'meaning' for one's existence—by responding to a perpetual escalation of self-impermanence anxiety and the drives for Leisure and Legacy needed to buffer it. The critical question then for biosocial management is: *How can we prescribe effective domains for Leisure and Legacy to 'quiet' our modern troubled minds, but at the same time equip these minds with sufficient focus and motivation to become champions in responding effectively to the converging catastrophes of the twenty-first century?*

It will be challenging but not impossible to design a greater selection of environment-friendly cultures for Leisure and for Legacy through personal accomplishment. An alternative motivational domain for legacy, with an emphasis on combining parenthood with religion—a 'breed hard–pray hard' type—is also conspicuous in modern culture (Aarssen and Crimi 2016). However, in the transition to (and beyond) the new and improved model for civilization (if the cultural evolution required to get there unfolds), one thing is certain: opportunity for legacy through parenthood will be much more limited than it is presently, or ever has been. This means that domains for leisure and personal accomplishment are likely to be especially vital and in high demand in the future. Some of these at least would be of great value in the same way that they have always been—e.g. cultures for legacy that involve hard work that also feels good because it is prosocial (Chap. 11), like volunteer work, charity, and commitment to egalitarianism. It will probably also mean that there will be greater demand for finding equanimity in religion and secular spiritualism (Lipka and Gecewicz 2017), and in therapies for 'escape from self' involving meditation, mindfulness, and cognitive behavioural therapy (e.g. Niemiec et al. 2010; Felder et al. 2014; Beck 2015), e.g. with a focus on 'rediscovering humility … a small, quiet, humble self' (Kesebir 2014). Perhaps the future will see successful societies where the daily routine includes a welcoming stroll through the community labyrinth (Zucker et al. 2016, https://labyrinthsociety.org/) (Box 13.3).

Box 13.3: A Labyrinth. (Image Source: Debra Dinda (https://www.flickr. com/photos/19576134@N00/32957075/))

Image source: Debra Dinda (https://www.flickr.com/photos/19576134@N00/32957075/)

Some are anticipating a future with more aggressive 'escape from self' therapies for calming the troubled mind—using genetic engineering and nanotechnology (e.g. Pearce 2015); or using 'digital world' technologies that provide 'immersive experiences' involving virtual reality, augmented reality, and avatars within meta-verse platforms (Vermes 2021); or by using new pharmaceutical prescriptions that incorporate psychedelics like psilocybin (Wong 2017; Davis et al. 2021).

As Carl Jung (1961) put it: 'We must not forget that for most people it means a great deal to assume that their lives will have an indefinite continuity beyond their present existence. They live more sensibly, feel better, and are more at peace'. Despite the absurdity in this assumption (viewed through the hard core lens of Darwinism), there seems to be something profound and insightful in Jung's plea here that I think resonates because it appeals to our intrinsic sense of empathy, and asks us—for the sake of equanimity and goodwill—to reflect on it with compassion for ourselves and for others (see also Asma 2018; Condon and Makransky 2020). Eagleton (2007) muses:

> We can divert our thoughts to the business of building life-giving mythologies—religion, humanism and the like . . . Such mythologies may not be true from a scientific viewpoint. But perhaps we have made too much of a fuss of scientific truth, assuming that it is the only brand of truth around ... such myths can be said to contain their own kind of truth, one which lies more in the consequences they produce than in the propositions they advance. If they allow us to act with a sense of value and purpose, then perhaps they are true enough to be going on with.

Perhaps other 'kinds of truth' measured in 'the consequences they produce', such as certain inspirations from the arts, may also have value for 'going on with'. For example, in the history of art, portraits that include the presence of a fly symbol-ize the transience of human life (Hickson 2020). One of the most cited scholars of the twentieth century, Noam Chomsky (1988) wrote: '... it is quite possible—over-whelmingly probable, one might guess—that we will always learn more about human life and human personality from novels than from scientific psychology'. As poet Ursula Le Guin (2016) describes it:

> Science describes accurately from outside, poetry describes accurately from inside. Science explicates, poetry implicates. ... We need the languages of both science and poetry to save us from merely stockpiling endless 'information' that fails to inform our ignorance or our irresponsibility ... by demonstrating and performing aesthetic order or beauty, poetry can move minds to the sense of fellowship that prevents careless usage and exploitation of our fellow beings, waste and cruelty.

Interestingly, Charles Darwin's grandfather, Erasmus Darwin, anticipated his grandson's theory of evolution involving natural selection before Charles was even born (Simon 2019)—in his poetry, published posthumously (*The Temple of Nature*, 1803):

> Organic life beneath the shoreless waves
> Was born and nurs'd in Ocean's pearly caves;
> First forms minute, unseen by spheric glass,
> Move on the mud, or pierce the watery mass;
> These, as successive generations bloom,

> New powers acquire, and larger limbs assume;
> Whence countless groups of vegetation spring,
> And breathing realms of fin and feet and wing.

Simon (2019) concludes: 'Literature anticipates and prepares; it can consider unrealised possibilities. Poetic fantasy prefigures scientific prose. If ever there was an era that required the scientific poetic imagination, it is ours. In an age that needs such healing and wisdom about the natural world, we too need poets who shall rise to the challenge, who can sing a song of evolution, of climate change. In some small but hopeful sense, maybe they can cure the world'.

Some people in old age (e.g. Boysen 2021), it seems, are more naturally equipped than many (e.g. Fingarette 2019) for finding reprieve, an inner peace, from the torment of self-impermanence anxiety. Going forward, however, I suggest most will inevitably find enduring solace from the time-honoured remedy that fired up and fueled the progress of civilization, marching relentlessly in every ancestral generation since the 'discovery of self' (Chap. 4), many thousands of years ago: if you can't 'keep calm and carry on', then *just keep busy* (Box 11.2). Because it is deeply ingrained in our evolved psychology, busyness, for most, may be the only effective distraction option for 'not seeing the infinite', and the only effective delusion option for 'connecting the finite with the infinite' (quoting from Tolstoy's (1872) reflection considered in Chap. 11). And if a 'keeping busy' culture can evolve centred on motivations and activities that are prosocial and pro-earth, with moral obligations to both contemporary and future generations of humanity, then perhaps this is also true enough to be going on with.

Coda

Authors of books like this—about mostly limitations, challenges, and other troubling prospects for humanity—usually try to conclude on a positive note. I have done some of this, earnestly, in this and the previous chapter. But in closing here, I wish I had more optimism to leave for the reader, and for myself. I am certain however about where I can continue looking and expecting to find it: with and from the leaders of tomorrow—our youth.

The scale of damage done to the earth's environment, its biosphere, and to the vitality of the human civilization project has multiplied several times within just my lifetime. My generation therefore is held accountable in large measure. But the culture of my youth was shaped by strong influence from the generation before it, delirious from the promise of oil and anxious to binge on the technologies, comforts, and toys that it gave us as the 'great acceleration' unfolded in the closing decades of the last century. The coming storm was just not on our radar. As Burr (2008) put it:

> In the years since the industrial revolution, we humans have been partying pretty hard. We've ransacked most of the earth for resources. A small part of the world's population

wound up with some nice goodies, but now we're … living off the natural capital of the planet, the principal not the interest. The soil, the sea, the forests, the rivers, and the protective atmospheric cover — all are being depleted. It was a grand binge, but the hangover is now upon us, and it will soon be throbbing.

My mission in this book has been to lay the foundation for a succinct and urgent conclusion: The converging catastrophes of the twenty-first century are products of our own biocultural evolution that has made us what we are—the only creature that refuses to be what it is. Accordingly, a promising future for our descendants will require that we teach, learn, and embrace quickly, not just the remedies that we know would work to minimize the impact of civilization collapse. It will also require that we teach, learn, and embrace effective and compatible remedies for regulating our frenetic drives—for legacy and leisure—that we must deploy in striving for untroubled minds, to convince ourselves that our existence is not absurd. We can do this, but the motivation and perseverance needed for this goal will require that we have a deep and broadly public understanding—presently lacking—of *how and why we evolved to become so driven*. With this, achieved through teaching especially our youth, I believe we will be equipped with the self-compassion needed to free our minds from the grip of runaway selection, and to stop and disembark from the runaway train of our imperilled civilization.

The cultural evolution of today's youth is immersed in the very visible and frightening signs of the approaching collapse. The future is now in their hands, and I am confident that— equipped with knowledge of what they are, as products of biocultural evolution—they will be prepared and ready for the task before them, determined to meet the challenge head on through social movements, activism, hard work, and wisdom. I believe it possible that the glory of their success awaits my children and grandchildren.

Earth, by the twenty second century, can be turned, if we so wish, into a permanent paradise for human beings, or at least the strong beginnings of one. We will do a lot more damage to ourselves and the rest of life along the way, but out of an ethic of simple decency to one another, the unrelenting application of reason, and acceptance of what we truly are, our dreams will finally come home to stay — Wilson (2012).

References

Aarssen L (2019) Dealing with the absurdity of human existence in the face of converging catastrophes. The Conversation, 1 May 2019. https://theconversation.com/dealing-with-the-absurdity-of-human-existence-in-the-face-of-converging-catastrophes-110261

Aarssen LW, Crimi L (2016) Legacy, leisure and the 'work hard – Play hard' hypothesis. Open Psychol J 9:7–24

Alexander S (2021) Searching for sanity in a world hell-bent on destruction. The Conversation, 13 May 2021. https://theconversation.com/friday-essay-searching-for-sanity-in-a-world-hell-bent-on-destruction-160447

Allen B (2019) From sacred to secular: Canada set to lose 9,000 churches, warns national heritage group. CBC News, 10 Mar 2019. https://www.cbc.ca/news/canada/losing-churches-canada-1.5046812

Asma T (2018) Why we need religion. Oxford University Press, New York

Beck J (2015) What good is thinking about death? The Atlantic, 28 May 2015. http://www.the-atlantic.com/health/archive/2015/05/what-good-is-thinking-about-death/394151/

Boysen I (2021) Love evolves and death isn't worth your worry – Life lessons from an 88-year-old. Aeon, 3 Aug 2021. https://aeon.co/videos/love-evolves-and-death-isnt-worth-your-worry-life-lessons-from-an-88-year-old

Brooks AC (2020) 'Success addicts' choose being special over being happy. The Atlantic, 30 July 2020. https://www.theatlantic.com/family/archive/2020/07/why-success-wont-make-you-happy/614731/

Burr C (2008) Culture quake: your children's real future. Trafford Publishing

Camus A ([1942] 1955) The myth of sisyphus, and other essays (O'Brien J, Trans). Knopf, New York

Chomsky N (1988) Language and problems of knowledge: the managua lectures. The MIT Press, Cambridge

Choron J (1964) Modern man and mortality. The Macmillan Company, New York

Condon P, Makransky J (2020) Modern mindfulness meditation has lost its beating communal heart. Aeon, 16 Sep 2020. https://psyche.co/ideas/modern-mindfulness-meditation-has-lost-its-beating-communal-heart

Curtin SC, Warner M, Hedegaard H (2016) Increase in suicide in the United States, 1999–2014. NCHS data brief, no 241. National Center for Health Statistics Hyattsville, MD

Davis AK et al (2021) Effects of psilocybin-assisted therapy on major depressive disorder: a randomized clinical trial. JAMA Psychiat 78:481–489

Eagleton T (2007) The meaning of life: a very short introduction. Oxford University Press, Oxford

Felder A, Aten HM, Neudeck JA, Shiomi-Chen J, Robbins BD (2014) Mindfulness at the heart of existential-phenomenology and humanistic psychology: A century of contemplation and elaboration. Humanist Psychol 42:6–23

Fingarette H (2019) A 97-year-old philosopher ponders life and death: 'what is the point?'. Aeon, 18 Feb 2019. https://aeon.co/videos/an-ageing-philosopher-returns-to-the-essential-question-what-is-the-point-of-it-all

Fleming A (2015) Why you want something even if it's bad for you: a rebel neuroscientist's research sheds new light on why desire is so powerful. Financial Review, 30 May 2015. https://www.afr.com/life-and-luxury/health-and-wellness/why-you-want-something-even-if-its-bad-for-you-20150527-ghamp3

Hazlitt W (1822) On the fear of death. In: Keynes G (ed) Table talk, essays on men and manners. 2010 Selected essays of William Hazlitt 1778 to 1830. Kessinger Publishing, Whitefish, MO. http://www.blupete.com/Literature/Essays/Hazlitt/TableTalk/FearDeath.htm

Hickson S (2020) Mike Pence's fly: from renaissance portraits to Salvador Dalí, artists used flies to make a point about appearances. The conversation, 9 Oct 2020. https://theconversation.com/mike-pences-fly-from-renaissance-portraits-to-salvador-dali-artists-used-flies-to-make-a-point-about-appearances-147815

Horowitz M (2018) The suicide wave. Medium, 25 June 2018. https://medium.com/s/radical-spirits/the-suicide-wave-f33f675e9ab8

Jung CG (1961) Memories, dreams, reflections. Random House, New York

Kesebir P (2014) A quiet ego quiets death anxiety: Humility as an existential anxiety buffer. J Pers Soc Psychol 106:610–623

Le Guin UK (2016) Late in the day: poems 2010–2014. PM Press, Oakland, CA

Lipka M, Gecewicz C (2017) More Americans now say they're spiritual but not religious. Pew Research Center, 6 Sep 2017. https://www.pewresearch.org/fact-tank/2017/09/06/more-americans-now-say-theyre-spiritual-but-not-religious/

McCallum J (2021) The tyranny of work. Aeon, 28 Jan 2021. https://aeon.co/essays/how-the-work-ethic-became-a-substitute-for-good-jobs

Moen RA, Pastor J, Cohen Y (1999) Antler growth and extinction of Irish elk. Evol Ecol Res 1:235–249

Mustafa N (2020) Long live leisure! It's time to make space for what we value in life. Ideas, 20 Feb 2020. https://www.cbc.ca/radio/ideas/long-live-leisure-it-s-time-to-make-space-for-what-we-value-in-life-1.5469745

Newnham W (1827) The principles of physical, intellectual, moral, and religious education, vol 2. J Hatchard and Son, London

Niemiec CP, Brown KW, Kashdan TB et al (2010) Being present in the face of existential threat: the role of trait mindfulness in reducing defensive responses to mortality salience. J Pers Soc Psychol 99:344–365

Nock MK (2016) Recent and needed advances in the understanding, prediction, and prevention of suicidal behavior. Anxiety Depress 33:460–463

Nugent C (2019) Terrified of climate change? You might have eco-anxiety. Time, 21 Nov 2019. https://time.com/5735388/climate-change-eco-anxiety/

Pascal B (1670) Pensées and other writings; Translation by Honor Levi (1995). Oxford University Press, New York

Pearce D (2015) The hedonistic imperative. David Pearce, Los Angeles. https://www.hedweb.com/welcome.htm

Pew Research Center (2012) "Nones" on the rise. Pew Research Centre, 9 Oct 2012. https://www.pewforum.org/2015/04/02/religious-projections-2010-2050/

Russell B (1957) Mysticism and Logic. Double Day, New York

Simon E (2019) How Erasmus Darwin's poetry prophesied evolutionary theory. Aeon, 29 May 2019. https://aeon.co/ideas/how-erasmus-darwins-poetry-prophesied-evolutionary-theory

Sterelny K (2012) The evolved apprentice: how evolution made humans unique. MIT Press, Cambridge

Thompson D (2019) Workism is making Americans miserable. The Atlantic, 24 Feb 2019. https://www.theatlantic.com/ideas/archive/2019/02/religion-workism-making-americans-miserable/583441/

Tolstoy L (1872) A confession. Unabridged Dover (2005) republication of the Aylmer Maude translation. Oxford University Press, London, 1921

United Nations (2019) One person dies by suicide every 40 seconds: new UN health agency report. 9 Sep 2019. https://news.un.org/en/story/2019/09/1045892

Vermes J (2021) Facebook to rebrand as a metaverse company. What is that? CBC, 24 Oct 2021. https://www.cbc.ca/radio/day6/introducing-the-metaverse-crisis-in-afghanistan-stuff-the-british-stole-islamic-influence-in-dune-and-more-1.6220405/facebook-to-rebrand-as-a-metaverse-company-what-is-that-1.6222140

Whyte C (2019) US suicide rate at its highest since the end of the second world war. New Scientist, 20 June 2019, https://www.newscientist.com/article/2207007-us-suicide-rate-at-its-highest-since-the-end-of-the-second-world-war/

Wilson EO (2012) The social conquest of earth. Liveright Publishing Corporation, New York

Wolfe S (2019) Millennials and the rise of hustle culture. Medium, 18 Feb 2019. https://medium.com/@samwoolfe_69260/millennials-and-the-rise-of-hustle-culture-2e1ba963f46b

Wong S (2017) Mind menders: how psychedelic drugs rebuild broken brains. New Scientist, 22 Nov 2017. https://www.newscientist.com/article/mg23631530-300-mind-menders-how-psychedelic-drugs-rebuild-broken-brains/

Zucker DM, Choi J, Cook MN, Brennan Croft J (2016) The effects of labyrinth walking in an academic library. J Libr Adm 56:957–973

Printed in the United States
by Baker & Taylor Publisher Services